D1130448

The Course of Industrial Decline

Johns Hopkins Studies
in the History of Technology

Merritt Roe Smith
series editor

LAURENCE F. GROSS

The Course of Industrial Decline

THE BOOTT COTTON MILLS OF LOWELL, MASSACHUSETTS, 1835–1955

THE JOHNS HOPKINS UNIVERSITY PRESS

BALTIMORE & LONDON

Printed in the United States of America on acid-free paper

The Johns Hopkins University Press
2715 North Charles Street
Baltimore, Maryland 21218-4319
The Johns Hopkins Press Ltd., London

Library of Congress Cataloging-in-Publication Data

Gross, Laurence F.
The course of industrial decline : the Boott Cotton Mills of Lowell, Massachusetts, 1835–1955 / Laurence F. Gross.
p. cm. — (Johns Hopkins studies in the history of technology; new ser., no. 15)
Includes bibliographical references (p.) and index.
ISBN 0-8018-4453-3
1. Boott Mills (Lowell, Mass.)—History. 2. Cotton textile industry—Massachusetts—Lowell—History. 3. Cotton manufacture—Massachusetts—Lowell—History. 4. Textile factories—Massachusetts—Lowell—History. I. Title. II. Series.
HD9879.B66G76 1993
338.7'677'0097444—dc20 92-25128

A catalog record for this book is available from the British Library.

TO THE MEMORY OF

MICHAEL BREWSTER FOLSOM

Contents

PART FOUR
Depression, Continuity, and Change, 1931–1946

PART FIVE
The Boott Mill Expires, 1947–1955

Illustrations

Acknowledgments

PEOPLE far too numerous to list, let alone remember, have offered me assistance and encouragement during the time spent preparing and writing this book. Many people in the cotton textile industry—owners, managers, and workers—have shared their understandings of its many facets. Librarians Martha Mayo, of Lowell University, and Clare Sheridan and her predecessor, Helena Wright, of the Museum of American Textile History, provided me steady assistance in the mining of their collections. Archivist Nicholas Olsberg and his assistants put the Flather Papers into coherent order (thus rendering them usable). While I was producing an earlier study of part of the Boott Cotton Mills complex for the National Park Service, David Zarowin provided irreplaceable assistance as a researcher. Thomas Leavitt and Paul Hudon were supportive and read an early draft. Thomas Dublin offered helpful commentary at a point when the manuscript still required much rewriting. Another Johns Hopkins's reader, Gary Kulik, offered a great deal of help, as did Philip Scranton and Gary Gerstle, with the penultimate draft. Merritt Roe Smith and editor Henry Y. K. Tom offered encouragement, and copy editor Ann Waters showed me where I had drifted from the path. I cannot properly express the degree of my debt, and appreciation, for the contribution of Marion Hall, who has been not only patient but also gracious as she typed, improved, and processed endless revisions.

I must also thank the Flather family, which helped keep the mill going, preserved its records, and made them accessible. A subvention from the Museum of American Textile History has been provided in the hope that the book might reach a wider audience. Finally, I want to express my gratitude to the many workers, most of them long gone, who perpetuated this enterprise through their skill and endurance.

Introduction

Given the acres of forests cut and barrels of ink spilled out of devotion to the recounting of the story of early Lowell, I have wondered about the advisability of attempting another account of the city. Henry A. Miles seemed to set the limits of such attention in 1845 in *Lowell, As It Was And As It Is,*[1] when he implied no need for consideration of "As It Will Be," or "How It Turns Out." Historians and others have been generally content with his concept of the bounds of the city's significance and have seldom strayed far beyond them.

I feared, of course, that the explanation for this phenomenon lay in the unimportance of events in the city after that time, or after the Civil War, a watershed in the country's history often used as a useful demarcation for the end of Lowell's originality and its significance for the larger concepts of the nation's development. On the contrary, however, I found that reasons abound to test the myths of early Lowell against its ultimate development, to follow through its story and determine how this novel enterprise evolved.

To ignore the city after its founding era, its so-called Golden Age of Yankee "mill girls," august visitors, and constant attention (both here and abroad) to the great experiment in the wilderness, is to slight not only the significance of its impact on the history of the United States but also the lives and contributions of the vastly larger group of inhabitants involved in its industry after the war than before, the hundreds of thousands who followed the tens of thousands who worked in the mills in the early decades. To follow industry's path in the city into the twentieth century provides the opportunity to compare performance with promise, measure results against predictions, and assess the role of an operation of Lowell's type in the national economy. Given the attention directed at the early history, the numbers of people involved in the later years, and the potential for insight into the nature and effects of the developments there, Lowell deserves no less.

I decided to follow the playing out of the story of the cotton textile industry in Lowell in terms of the rise and fall of one corporation, founded as the Boott Cotton Mills in 1835, reorganized as the Boott Mills in 1905, under which name it continued until it closed in 1955. The reasons for such a single-minded approach are several. They parallel, in part, the attitude Thomas Dublin displays in *Women at Work,* focusing on one community in order "to gain access to those human experiences that tend to disappear among the larger generalizations and aggregates of economic history."[2] By concentrating on one corporation, I hoped to achieve a new level of attention to the human element in the story and at the same time to make the nature of the story itself more meaningful than it becomes when seen only in terms of gross statistics. Following the specific story should also help make clear the implications of the general developments it represents.

Alfred D. Chandler, Jr., noted the scarcity of such an approach in *The Visible Hand* and at the same time offered an explanation:

> Historians have been attracted by entrepreneurs, but they have rarely looked closely at the new institution these entrepreneurs created, at how it was managed, what functions it carried out, and how the enterprise continued to compete and grow after the founders had left the scene. Instead they have argued as to whether these founding fathers were robber barons or industrial statesmen, that is, bad fellows or good fellows. . . . Few historians, however, have tried to trace the story of a single institution from its beginnings to its full growth. None have written about the rise of modern business enterprise and the brand of managerial capitalism that accompanied it.[3]

I proposed to write about an embodiment of these particulars, a single institution, its operations, functions, and performance, and the roles of various people in accomplishing them. Such concentration will permit, I hope, a clearer understanding of the overall picture in which this American corporation played its part and allow attention to the functions of the workers, the managers, and the owners and something of the relationships between all of these elements and the political and social arrangements within which they acted, all of which Chandler consciously omits from his book. Following the successes and failures of one of the Lowell cotton mills permits an understanding of the nature of all the corporations' operation, their relationship to the national economy, and their effects on those who worked for them and lived with them.

The attention given to early Lowell created a set of beliefs about the nature of the developments it initiated, and these beliefs created assumptions about their operation over time. Close inspection of the Boott should enable us to determine the merit of the original assessments and to evaluate the accuracy with which they have been used by later writers. These impressions and evaluations also affect perception of the work and the workers. Most people would agree that Lowell developed a new type of labor for those who went into the mills in the early years, as expressed by David Brody:

> The factory worker became the servant of the machine that performed the actual work or, if he did the endlessly repetitive jobs on the assembly line, he became a part of the machine. . . . For the factory worker, there was nothing tangible at the end of the day's work to which he can point with pride and say, "I did that—it is the result of my own skill and my own effort."[4]

Such changes came early but incompletely to textiles, where workers were certainly made servants of the machines, but where they did make a visible product in many cases, whether roving, yarn, or cloth. I have wondered what part of the equation primarily shapes evaluation of work by participants and historians: pride in production, amounts of skill, or perhaps outside recognition of it? And what combination of these elements operates on various workers at various times to shape their feelings about what they do, and to shape historians' opinions about the foreign tasks of those about whom they write. Again, attention to a specific operation and its conduct may permit a greater than usual degree of comprehension of the nature of the work, the demands it made on the operatives and their responses to them. If there is room for doubt on something as axiomatic as the low skill requirements of machine-tenders in the cotton textile industry, then what other areas of assumption may offer fruitful ground for investigation?

Throughout this account, the relationship between the components making up cotton mill technology—structures, machinery, workers, and management—will appear prominently. The intricacies and evolution of their interactions offer insight into the intentions and outlooks of the people involved and their different perspectives. The ways in which these elements developed and their relationships changed run through the manuscript as from novelty to maturity.

The relationship of labor to industry receives particular attention. The nature of the corporate paternalism so often associated with Lowell's early days deserves scrutiny. Linus Child, agent for the Boott Cotton Mills, 1851–52, was charged with threatening the discharge of any worker who voted "contrary to the interests of the corporation."[5] James Green charges that the "corporate paternalism" characteristic of the city's early years later changed to a "corporate totalitarianism that effectively checked union organization," but how does one square the idea of change in this concept with the evidence of consistent conduct such as that of Child (who represents but one example)?[6] In this context, I will include the record of the Boott with regard to its boarding houses and examine when the discipline associated with them disappeared.[7]

That there was little labor organization in Lowell over the decades cannot be questioned, but the causes of the inactivity invite various explanations. Was it prevented by the power cited by Green, a respect for management competence like that noted in Philadelphia by Philip Scranton, or by the effect of organization on the part of the managements, or by the still more powerful force of management ideology described by Brody: "American workers might engage in pressure tactics, but, as Sumner Slichter remarked, they knew they were breaking the rules. It would be hard to imagine a more insidious check on so fundamental a phenomenon as the self-activity of the work group."[8] Such lines of inquiry promise useful investigation not only for the positive answers they may provide, but also for the ways in which workers determined that organization and protest were, ultimately, justified.

In studying a single Lowell corporation, and particularly its conduct after the Civil War, I will emphasize issues associated with the later period, as well as those which bridge the earlier and the later. In Constance McLaughlin Greene's words, "Massachusetts industrialism was soon to embark upon a career of activity which should make the preceding era seem primitively limited."[9] I would argue that it is in the course of industry in this post–Civil War period that its nature and effects stand most clearly revealed.

Just as many important issues were not resolved in the early years, others were not raised. The contest between the North and South to provide a home for the cotton industry had not begun, and it is part of the neglect of the later period that enables Scranton to write that "the precise mechanism of the decline (of the industry in the North) has not been specified."[10] Given the pivotal nature of the role of runaways in current discussions of industry, such an omission regarding textiles' origination of corporate flight seems remarkable.

While no one should claim to offer too many answers, I hope that this study of a given enterprise will offer insight into broad questions and will promote an understanding of issues not fully developed in more general studies. By looking carefully at the conflicts and tensions as well as the accomplishments of the various parties in one operation, we may come to understand more clearly the relationships between worker and owner, corporation and community, and family and economy which shape our society.

The first section of this book extends from the textile industry's origins through the Civil War. While of necessity broader and more general than those to follow, Chapter 1 establishes some of the procedures and attitudes of Lowell's early years. Chapter 2 covers the period 1850–70, describing conflict among owners, some practices of mill management, and the approaches they embodied. The next three chapters extend from 1871 to 1904, the era of Lowell's great expansion. In Chapter 3, records of the construction and equipping of buildings offer detailed views of this activity; in Chapter 4, the agent's correspondence illuminates the role and attitudes of this important player. The Boott's failure at the end of the period vividly reveals the nature of the operation, as we see in Chapter 5. In Chapters 6–8, which cover the years 1905–30, the corporation arises, phoenixlike (I can't resist saying), from its ashes. The sources of its reborn success lead to decisions that seal its fate by the period's end, at the start of the Great Depression. Chapter 9 demonstrates the operation's capacity to persevere and even prosper as run by its new owners, and Chapter 10 considers the changing role of government and that of a new element, the Textile Workers' Union, American Federation of Labor. The last section, Chapters 11 and 12, spans the shortest time, 1947–55, yet it brings the story to a close and considers the roles and views of all the participants. Finally, the Postscript looks back over this history and discusses its portent in current economic climes. Different elements of the history receive more prominence in different chapters. Over the course of the narrative, I hope they combine to present a full description of the Boott Cotton Mills and its populace.

Part One

Lowell through the Civil War

Chapter One

Lowell to 1850

Precursors

The successful production of cotton yarn on powered machinery organized by Samuel Slater, in Pawtucket, Rhode Island, in 1790, signaled the reordering of textile production in the direction of modern industry. He achieved novelty in constructing and organizing equipment in this country and in manufacturing for investors' profit. He remained allied with past practice in that he himself worked in production and served hand-weaving operations. The role played by an owner in the factory, the mixed modes of production, and their small scale at first obscured the pattern and meaning of the alterations in the economy and in human relations which Slater's new methods initiated.

Fuller evolution of the potential of Slater's Mill occurred not at Pawtucket, but at Waltham, Massachusetts. In the decades leading up to and following 1800, the major fortunes in New England had been amassed through maritime trade, a mercantilist capitalism. Merchants found that path thwarted by the British during and after the War of 1812, which assured English domination of the seas. Francis Cabot Lowell, a leading trader, had visited Great Britain in 1810 and seen the future, he believed, embodied in powered spinning *and powered weaving* of cotton cloth. Lowell and some of his associates hired an exceptional machinist/machine-builder/inventor, Paul Moody, and began work on a project to reinvent the power loom, not yet available in this country and unavailable from the protectionist English. They also formed the Boston Manufacturing Company (BMC) to carry out their plans.

Once they had produced a workable loom, they established the first integrated textile mill in the country, on the Charles River in Waltham. Raw cotton entered at one end, and cloth came out the other. The organization of the new enterprise as a joint-stock company was equally novel. The owners were financiers, not cotton manufacturers, and they administered the factory from their Boston offices. Their combination of powered equipment, factory organization, and financial acumen proved very successful.

The BMC approach showed careful planning and thorough rationalism. The headings in its ledger book listed the elements to be coordinated: Land and Water Power, Buildings, Machinery, Labor, and Administrative Supervision. As George Sweet Gibb put it, "How successfully these factors could be combined by unpracticed men to turn out an unfamiliar product on equipment not yet invented was far from clear."[1] In a factory to be peopled with workers neither present nor trained, he might have continued. In any case, the ledger indicated that labor simply represented a commodity to be managed.

The BMC intended to pursue a particular style of powered production of cloth at Waltham. They planned to reduce every aspect of cloth production to its simplest elements and assign just one element to each worker. Every effort aimed to minimize the amount of skill required of a given operative. For example, they carefully chose a line of production permitting utilization of the throstle spinner rather than the spinning mule in order to avoid the skilled, and sometimes organized, mulespinners. Labor was not to intrude through skill to interfere with the investors' control over production. Moody facilitated their scheme as he improved numerous machines and developed "stop-motions" to arrest the action when one malfunctioned or needed attention.

Soon the mills at Waltham produced thirty miles of cloth a day, but the power of the Charles River would support no further growth. The factory had generated impressive profits and the owners hungered for expansion.

FOUNDING THE NEW CITY

In response, the Boston investors "found" the Pawtucket Falls of the Merrimack River at East Chelmsford. The thirty-foot fall in the mighty Merrimack could power ninety mills the size of Waltham's.* It is exemplary of the power, the vision, and the potential of the BMC that this great source of energy had not previously been seen to have value as the property of any individual, for none could undertake to harness it. To the BMC, however, it represented the factor by which to multiply their earnings at Waltham.

*Decades earlier the town had given land to a handweaver in order to assure the community the availability of his skill. A few farmers occupied the land adjacent to the river, and several small powered operations, most using wing-dams capturing a small fraction of the river's force, operated in the vicinity of the falls and on the nearby Concord River (which entered the Merrimack a mile below the falls). Wing-dams blocked and diverted into a waterwheel only a small part of the river's flow. They required neither capital nor expertise on the scale of that necessary to dam the entire river.

An earlier corporation, The Proprietors of Locks and Canals on the Merrimack River (Locks and Canals), had built the Pawtucket Canal for transportation around the falls in 1792–96. Displaced by the Middlesex Canal, it provided no obstacle to acquisition by the BMC: the two groups were friends and relatives of one another. Similarly, local farmers provided no serious impediment to the development plans: "So unconscious was America of the value of its water power that [Patrick Tracy] Jackson with very little difficulty succeeded in acquiring a majority of the stock of the Locks and Canals Company and title to the key farmlands."[2] Jackson and the other BMC investors hired Kirk Boott, son of another Boston mercantilist, to buy the necessary lands and, ultimately, to manage the operations erected there. Because of the public's ignorance of both the land's potential value and of the BMC's plans, land which a year later would be valued at $4,300 could be had for $20 an acre. Only Thomas Hurd, a local manufacturer making woolen goods in a mill on the Concord, managed, through his knowledge and the felicity of overhearing a conversation among the principals in Boston, to obtain a substantial price, some $2,100, for land important to the project.

Development of Lowell, named to honor the deceased advocate of the plans, proceeded rapidly. With the imperious Boott in charge on the scene, Irish and Yankee laborers repaired the dilapidated Pawtucket Canal for waterpower use and dug a new one, the Merrimack, to bring water to the Merrimack Manufacturing Company. Boott's military engineering education (Sandhurst) combined with his aristocratic airs to produce disdain for (and from) the workers and residents alike, and he exhibited single-minded devotion to his employers. Only the latter thought well of him. By 1823 cloth was being produced by this giant new mill complex, dwarfing its Waltham progenitor. Quickly thereafter the same group of investors established more companies in Lowell on the same model: the Hamilton (1825), the Appleton (1828), the Lowell (1828), the Middlesex (1830), the Suffolk and the Tremont (1832), and the Lawrence (1833). The Boott Cotton Mills followed in 1835, and then the last of the group's major corporations there, the Massachusetts (1840).

These corporations (only the individual buildings were called mills) greatly resembled one another. They were closely held by a small group of investors, membership in which overlapped substantially from one to another; they produced a narrow range of cloth of a comparatively coarse type that required as little skill as possible in production; they intended, by keeping their products discrete, to minimize competition with one another. They expected to attract a native-born female workforce from the surrounding countryside to tend the machines under male supervision. Long seen as the forward-looking development of the economically adventuresome, Lowell has recently been revealed as the cautious development of the socially conservative. Robert F. Dalzell, Jr.'s *Enterprising Elite: The Boston Associates and the World They Made* does nothing to diminish the significance of their industrial activity, but it recasts their motives and makes visible the nature of their management of the new industry. Basically, Dalzell shows that the Boston Associates used textiles as part of economic, political, and social efforts to perpetuate their families and class in ways closely attached to the past. The system represented conservative planning dedicated to the security of the original investors, not to continuing growth and certainly not to profit which entailed risk. Furthermore, the system was to operate without the direct interference of the principals, who would thus be permitted the time to play the formative or controlling roles they coveted in society: "Far from the production of wealth in the usual sense, the goal was the preservation of fortunes already made, positions already won." The intention, in fact, was to restrict and contain the economy in the interest of the preservation, the security, of the Boston Associates.[3]

LAUNCHING THE BOOTT COTTON MILLS

The Boott Cotton Mills present an excellent example of the financial side of this situation. Its incorporators were three: Abbott Lawrence, also significant in the Lowell Machine Shop, and, soon, the Atlantic Cotton Mills and Pacific Cotton Mills of Lawrence, Massachusetts; Nathan Appleton, an original member of the BMC, prominent in innumerable textile enterprises, largest owner of the Hamilton, and later a congressman; and John Amory Lowell, nephew and son-in-law of Francis Cabot Lowell, trained under Kirk Boott, Sr., and partner of John Wright Boott (brother of the younger agent Kirk), before becoming treasurer of the new corporation. Like many of those involved, Lowell was young, just thirty-six when he assumed the most important position in the management of the Boott Cotton Mills. The original indenture reveals the close relationship between Locks and Canals, the company set up to develop the land and water rights, and the new corporation. Six directors signed the indenture for each firm, but only eight people were required to perform the deed; this was not an insoluble algebra problem—George W. Lyman, Henry Cabot, Kirk Boott, and J. A. Lowell signed as directors of both firms. The Boott paid $84,020 to Locks and Canals for land and "nine mill privileges" on the lower level of the canal system. The yearly rent for the waterpower, "sixteen ounces, seventeen pennyweights, twelve grains troy weight of gold, or two hundred and sixty ounces troy weight of silver," would simply pass from one pocket of these men to another.[4]

The corporation was named for Kirk Boott. Since his house stood on the site and had to be moved to make way for the mills, perhaps this action was an effort to mollify the haughty Boott, but more likely it was intended to honor the man whose service to the financiers endeared him to them as much as it alienated him from the local people. Regardless, he passed away before the operation started up.

To house the operation, the Boott built four mills between 1836 and 1839. They stood four stories high on the south, or millyard, side, five on the north, facing the river. Small Picker Houses, separate because of fire danger, were appended to each. The mills were not identical; for example, #1 used cast-iron pillars, #2, wooden. All the Lowell mills were built of brick made from the plentiful clay of the region. It has often been suggested that brick was used to facilitate planned later connection of the structures into larger facilities, but floors mismatched in levels by more than a foot at the Boott indicate the absence of such planning.[5] A Counting House stood in front of the mills, adjacent to the Eastern Canal. Three stories with an attic, it provided space for administrative and accounting functions, as well as serving as a dormitory for the night watchmen.[6] Each mill contained all the equipment needed to turn raw cotton into cloth and operated as an independent unit. The Locks and Canals machine shop supplied machinery for three of the mills between 1836 and 1839 for a total cost of $174,000.[7] The complex grew in 1847–48 with the addition of mill #5, which added 120,000 sq ft, about the size of two of the previous mills. A central Picker House went up about 1860.[8] These small, narrow buildings shared more with barn construction of the time than with the sort of "modern" mill construction developed during the final quarter of the century. Separate buildings, they altered the landscape but did not wall off the manufacturing city from views of the countryside across the Merrimack.

The Eastern Canal, flowing between the mills and the city, presented a placid picture which belied its role of delivering the water that would power all the machinery in the mills. The corporation moved immediately to utilize four of its millpowers. In a wheelpit in the center of the basement of each mill turned a breast wheel about 60' long and 17' high, or the amount of fall between the canal and the river: "These breast wheels were of simple construction, consisting of 30 or more buckets on the periphery of the wooden cylinder, which were filled from a gate at the end of the penstock, and the waste of water being prevented by a breast or apron back of and close to the outer edge of the buckets." Each wheel could generate about 60 HP, or one millpower—that is, energy sufficient to operate 3,584 spindles running on #14 yarn, and all the other equipment needed to turn cotton into cloth—a measurement of power defined by the second mill in Waltham. In other words, each of these mills roughly equaled that one.[9]

The four wheels filled the mills' needs until the mid-1840s, when James B. Francis, Locks and Canals agent, introduced his waterpower expansion program. The Lowell corporations then took control of and regulated the flow of New Hampshire's northern lakes (Winnipesaukee, Squam, and Newfound) and added the Northern Canal to increase the water flow into the system. The Boott added two center-vent turbines of Francis's design, made by the Lowell Machine Shop (LMS) and generating about 230 HP (or nearly four times the power of a breast wheel) each. Francis's expansion of the system increased the Boott's available power from 10 to 17 26/30 millpowers, and his turbine design helped them utilize it more efficiently.[10]

The new mills and wheels would produce "drillings, sheetings, shirtings, linens, fancy dress goods, and yarns. The looms used were the latest improved machines built under the direction of the proprietors, who were expert machinists."[11] (Who "employed" expert machinists would have been more accurate.)

The Lowell corporations had begun by adopting cam looms, as had Waltham, rather than those of the crank, or Scotch, type, which were to prevail, and their efforts must have been somewhat hampered by that mischoice. The loom episode was testimony to both their capacity to err and their ability to survive (as well as gloss over) their mistakes through their great economic power. The above claim ("latest improved machines") should have indicated that the Boott started operation with the crank loom that was introduced to this country by William Gilmore at about the same time as the Waltham cam loom. Gilmore's work in Rhode Island did not receive the publicity of the path followed by the innovators and publicists to the north. Lowell eventually had to shift to the clearly superior crank loom. The preeminence of the Waltham loom in historical accounts, despite its early obsolescence and lack of priority of invention, offers profound evidence of the self-promotional nature of the BMC's endeavor and of its success.[12]

Despite the best efforts of the incorporators, the new operation could not entirely avoid competing with the others. In May 1836, Amos Lawrence, in Boston, received a report from Robert Means, at the Suffolk Mills: "We have a superabundance of water—girls and men plenty—and every thing will conspire to occasion great work until about July and August, when the demand for girls to supply the Boott Mills will probably sensibly affect us."[13] The issue of help, its source, nature, availability, training, and control, stood in the minds of all concerned as a major aspect of these new operations.

LABOR IN THE NEW MILLS

No labor force stood trained and ready for the work of making cloth with machinery at Lowell. Worse, the bad press of the English textile factories, the "dark Satanic mills" of the romantic poets, had prejudiced U.S. residents against the idea of factory work. The public relations efforts of the BMC were intended, in part, to overcome this feeling.

As part of their plan of operation, the new corporations looked to obtain workers not by paying sufficient wages to draw males away from their occupations (from which they might also bring ideas about workers' perquisites and rights), but to attract a less skilled workforce of rural female labor. Domestic service, the only widely available alternative employment paying cash wages, was repugnant to many women because of its subservient nature. Farm life had taught them how to work hard and not to fear hard work. If they could be attracted, they could do the work.

In 1835 Andrew Ure described the monumental difficulty facing the new industrialists. He denigrated the accomplishment of inventing the machinery for cotton manufacture in contrast to that of overcoming "desultory habits of work" and creating through human labor the "unvarying regularity of the complex automaton," the factory:

> To devise and administer a successful code of factory discipline, suited to the necessities of factory diligence, was the Herculean enterprise, the noble achievement of Arkwright. Even at the present day, when the system is perfectly organized, and its labor lightened to the utmost, it is found nearly impossible to convert persons past the age of puberty, whether drawn from rural or from handicraft occupations, into useful factory hands. After struggling for a while to conquer their listless or restive habits, they either renounce the employment spontaneously, or are dismissed by the overlookers on account of inattention.[14]

The owners' needs required new attitudes to govern human relations, and only the establishment of a new work discipline could provide them. The task, then, was twofold: to attract workers and to teach them a new way of work, a new way of life. Responses to the two needs dovetailed, with many measures serving both ends.

The various factors leading the workers to the mills are difficult to weigh. Escape from rural isolation, family domination, and farm drudgery perhaps all played roles, as did opportunities for the company of contemporaries, intellectual stimulation, travel, marketable skills, and cash wages. Furthermore, the mills may have been seen to offer the benefit of eliminating a laborious, monotonous task from their lives, that of producing cloth by hand. In case these or other elements acted insufficiently on the women, the mills jointly hired recruiters to travel through the hinterlands and encourage migration. Through all of these means and more, the owners were able to draw the women to their new city. Through a strict system of supervision of the women's every hour, working and free, waking and sleeping, they could overcome the reservations they and their parents might hold about factory life. They provided boarding houses and matrons, their cost deducted from the workers' pay. The isolated and previously undeveloped location gave an opportunity to create and then control the social matrix within which the factories operated.

The role of labor in Lowell reflected its part in an operation governed by economic calculation. Its treatment and use were in every way purposeful. In the words of Nathan Appleton, company housing "secures an excellent class of work[ing] people which tells materially in the prosperity of the company." He saw the entire effort as an entity, a construct of the owners, not a scene of human endeavor on the part of the local participants: "We are building a large machine I hope at Chelmsford."[15] Those ledger headings again, as housing, labor, power, buildings, et al. stood together as a construct of the BMC. Every aspect was to make its contribution, with the result seen as mechanical in toto as the looms themselves.

Henry Miles, prominent among the apologists for the investors, described the logic behind this point of view:

> The productiveness of these works depends upon one primary and indispensable condition—the existence of an industrious, sober, orderly, and moral class of operatives. Without this, the mills in Lowell would be worthless. Profits would be absorbed by cases of irregularity, carelessness, and neglect; while the existence of any great moral exposure in Lowell would cut off the supply of help from the virtuous homesteads of the country. *Public morals and private interests, identical in all places,* are here seen to be linked together in an indissoluble connection. Accordingly, the sagacity of self-interest, as well as more disinterested consideration, has led to the adoption of a *strict system of moral police.*[16] [emphasis added]

Miles' rationalization of this system of social control demonstrated the hubris and the pervasive self-justification of New England's financial powers; it left no room for "disinterested considerations." The arrogance of ownership appeared in its ability to define and regulate both "public morals and private interests" for its own advantage. A great distance separated it from those to whom it could assign habits of "irregularity, carelessness, and neglect," in conflict with the evidence of generations of hard work and success across New England. One wonders, in fact, how the sons and daughters of the "virtuous households" could tend so powerfully, upon their arrival in Lowell, toward such base habits.

Similarly, a Lowell mill-owner spoke frankly about the reasons for and utility of providing schooling. Educated help produced more, exhibited "better morals," and was "more ready to comply with the wholesome and necessary regulation" of the mill, even at times of wage-cuts, while "the ignorant and uneducated I have generally found the most turbulent and troublesome, acting under the impulse of excited passion and jealousy."[17] Once the rationale for the industry was based on profit rather than individual or social need, it needed new techniques to motivate workers. Educating them to believe in a supposed identity of goals helped predispose them to accept a discipline serving capital's interests. Thus, as Philip Scranton points out, the capacity to reason is tied to letters: those without schooling can't reason and are thus not quite human. It was indeed a new world being created.[18]

Owners saw the workers not as people like themselves, but as a commodity to be managed; workers who expected the investors to consider them as somewhat equal, fellow Yankees, were naive. A letter from Boott agent B. F. French to William Boott in 1844 indicated a willingness to exploit workers: "The evils which constant employment and want of amusements are calculated to produce, if persisted in too long, are to a very great extent counteracted by periodical visits to their friends." While willing to expose the women to "evils," the owners also had other reasons to encourage turnover. Miles noted the importance of homes elsewhere to which workers could return in slow times, and thereby "not sink down here a helpless caste, clamouring for work, starving unless employed, and hence ready for a riot, for the destruction of property, and repeating here the scenes enacted in the manufacturing villages of England."[19] While schools taught compliance, experience would be educating the workers in the reasons why they might rebel, he implied. Miles admitted the absence of any identity of interests between owners and workers and only hoped to avoid disruptions through turnover. Factory work, he expected, would teach the children of "virtuous homesteads" that they had no right to expect any return beyond an hourly wage when they were of service. Previously, corporations such as towns had responsibility not only for their own success, but also for the well-being of their members. In extreme cases, that meant poorhouses or public work. The new corporation devoted itself entirely to financial success, denied reciprocal responsibility to its employees, and left their care when unneeded to their families or public agencies. As a citizen lost rights by crossing political boundaries, so an employee lost them by leaving work, regardless of the impetus.[20]

Recognition of disparate interests was widespread. The owners found that not even the comparatively well-treated overseers could be expected to direct their energies for the company's benefit without special inducement. As early as the 1830s, supervisors began to receive premiums keyed to production, in effect piecework, to induce them to push the work at top speed. For some, attachment to old concepts of cooperation among workers rather than identification with the interests of others was proving a hard concept to abandon.

The new manner of production and economic organization also produced revolutionary change in the nature of work itself. For the first time people worked at a pace set relentlessly by machines. As spools whirled and lays beat, someone had to service the spinning frames and looms continuously. An element of freedom previously available was lost to these workers.

Owners aimed to segregate tasks until they could be learned quickly, to diminish the capability of the employees to bargain over wages or conditions, or to affect production. The machine-tenders were to be easily replaceable. The intention, and result, was to develop "labor-cheapening" machinery, cost-saving not labor-saving. The Lowell system sought and incorporated every technique by which it could reduce the workers' importance (skill, independence, responsibility, creativity) and cost. They stuck to the production of comparatively coarse cloth (leaving the finer grades to such alternate centers as Rhode Island and Philadelphia), which fit their rather crude machinery and their general approach. They escaped the need for the skilled help associated with fine goods or woolens (e.g., mulespinners, cloth finishers). The emphasis on stop-motions exemplified this intent. These mechanical devices stopped a machine in the case of malfunction or lack of raw material, just as an attentive operator would. Yet the common portrayal of the women employees of Lowell, the famous "factory girls," as unskilled operatives deserves more careful consideration than it has traditionally received. The young women of the region had abilities and training which made them well-suited for work in textile factories. They were accustomed to hard work and familiar with mechanical operations seen at the pervasive carding, saw, grist, and fulling mills, and numerous smaller textile mills which dotted the area. As part of a domestic production unit likely involved in preparing and spinning fiber, making and repairing garments, they had developed a full range of abilities related to handling fiber and thread and were prepared to adapt to and be adapted by their experience in Lowell's manufactories. Because these skills were common among them, they did not lead readily to high wages nor create a strong bargaining position. Since owners wanted wages to remain low, they encouraged the definition of the women's jobs as unskilled. However, a look at some of the tasks expected of a weaver, for example, offers another point of view.

A weaver, a woman at this time, operated two or three looms. Maintaining the filling supply represented the primary limiting factor. Each time she observed a shuttle carrying a nearly empty bobbin, she had to stop the loom in the proper position, move the lay to free the shuttle, remove and replace it, and restart the loom. Careful procedure would prevent unevenness in the cloth. She also had to adjust warp let-off and cloth take-up as the size of these two rolls varied (in inverse proportion), detect and repair broken warp threads before they led to escalating damage to the product, detect broken filling threads, and generally assure successful operation of the loom, calling a fixer when necessary. Each task required attentiveness; most required skills, both of judgment and execution. Adjustments were made by evaluation, not measurement. Warp repair required a particular knot, reinsertion of a thread into the correct heddle in the correct harness and the proper dent in the reed. Failure in any aspect of any chore produced defective cloth, marketable only at a sacrifice. Given piecework wages, speed was of the essence. Only those who did not (or do not) perform such work would be likely to describe it as unskilled. In fact, a now-anonymous writer in the Boott's Letterbook in 1859 recognized the degree to which skills increased in the mill. Addressing the importance of worker persistence, he wrote, "They will be worth more to us the last six months than they are the first twelve" if new workers could be convinced to stay. Doubling their value after a year's experience implied a considerable period of learning.[21]

To describe these jobs as unskilled is simply to devalue dexterity, quickness, keen eyesight, and substantial grace under pressure. Given the female origins of such jobs in this country, this devaluation comes as no surprise. Parallel skills of male handweavers previous to power weaving had not been similarly denigrated. Furthermore, the assessments which worked to devalue these skills were those of the men who paid for them, a self-serving judgment.

As the system became more firmly established and the workers more acculturated, the machine operation and flow smoother, the managers raised the number of machines per employee, the "stretch-out," as well as the machines' operating speeds, the "speed-up." At the same time, piece rates, and total pay, went down.

RESPONSES TO LOWELL

Of the people who populated these mills we have less detailed knowledge than of those who controlled them. French reported that in 1844 the 816 women employed at the Boott included 40 "supposed to be Irish," and just 43 illiterates, implying a well-educated native-born assemblage of women with English heritage.[22] Rather than try to stretch statistics, however, we can describe the introduction and experience of a few individuals at the Boott, based on a rare surviving series of letters. In the context of limited evidence regarding such experience, their similarity to other sources gives them the appearance, at least, of being representative.

The letters involve the Metcalf family of Winthrop, Maine, and their descriptions of events and beliefs in the first half of the century, particularly the 1840s. Despite the town's isolated location (near Augusta), in 1819 a letter cited the operation at a local cotton mill of twelve "water looms."[23] Even at that distance from the centers of population, industry was making itself familiar to the inhabitants. Living on a farm or in a village did not mean ignorance of the new equipment.

In 1843, Charles and Sarah Metcalf left Winthrop for the "city of spindles," as they called Lowell. Charles described a trip by several stage coaches of various sizes and degrees of crowdedness which carried them as far as Portland, where they stayed overnight and boarded a train. To that point the endeavor had meant constant motion-sickness, as feared, for his sister Sarah, as well as the tribulations of snow storms and deep mud for all. Still, they knew that once they reached their destination they would find friends and relatives with whom to visit, comrades and kin who could help them find work and train them once in the mill (Dublin found that two-thirds of the Yankee women at the Hamilton had relatives there).

Work retained an informality at this time, for they wrote of visiting their relatives Lavinia and Caroline in the Boott; of the former they noted, "Lavinia looks good but has not got looms yet." They expressed a fear of the adjustment to factory work, as well as the problem of passing through training into productive and financially rewarding employment. After a day or two of visiting family, Sarah joined her relatives in the Boott: "Monday she went into the mill, Mr. Wood's [apparently known to the family] room. She has not got looms yet but hopes to have soon. The overseer is full of help; at present Sarah [Caroline?] tends 3 looms next to Lavinia." Sarah made $6.12 1/2 plus board as a helper for her first four weeks and still hoped for looms of her own "in a week or two." Joining relatives and working under the oversight of someone familiar to the family must have eased the entry and the training. The three-loom load indicated a 50-percent increase (stretch-out) over that of 1838.[24]

Despite the factors facilitating the move, Charles quickly recognized the distance between Winthrop and Lowell and the ways of life associated with each. Regarding the possible migration of sister Mary to Lowell, he wrote that she might find it a difficult adjustment. He noted, "Every one here is wrapped up in self and looks out for No. 1." It seems a very modern place.

Charles found work at the neighboring Massachusetts Mills, where he worked carrying roving on bobbins in the Spinning Room for 63 cents a day. Charles described his job as "rather hard work. . . . I have to carry out and lay on top of the frames 5,600 or more bobbins every day, and the same number of empty ones back again so you see I have to be pretty lively." He ascribed his daily nosebleeds to the "fly" (lint) in the air. Exhausted, he was embarrassed because he fell asleep in church, a fact he blamed not only on his condition but also on the fact that the congregation there didn't rise at prayer times, as he was accustomed to, but only for singing at the beginning and end of the service. Life in the city required numerous adjustments. Unlike the women, Charles could, and did, hope for an escape from the mill to the much better and more rewarding work at the "big shop," the Lowell Machine Shop, where he had another relative.

When sister Mary arrived, Mr. Wood couldn't "use her yet." Both Sarah and Lavinia believe "she will make a better spinner" than weaver, indicative of the differentiation that persisted among the factory jobs.[25] Mary's letters demonstrated ways in which she differed from the others, helping explain Charles's reservations about her coming and the others' feeling about her greater suitability for spinning. In November 1847, she reported that she worked for Mr. Gaye in No. 1 Boott Winding, which she "likes very much. . . . I get out of the mill between 3 and 4 on Saturday." After a desperately lonesome Thanksgiving and another month of work, however, she wrote that she hoped to move to a job on winding in the attic of Mill #4: "I do not think my eyes will stand it to light up this winter and in the attic we shall not have to light up much if any." The dormer-windowed attic space apparently offered better lighting than the wider spaces of the lower floors. Work by the artificial lighting of oil lamps was both hard on the eyes and smoky.

Mary's letters suggested reasons for her going to Lowell and staying there. She wrote with exasperation: "How could you think mother that I was going to do house work when you heard me positively say I would not do it." She also felt, and objected to, her mother's long-distance attempts at bossing her.

At the same time, her adaptation to Lowell had been incomplete not only within the factory, but also in relation to its management. She referred to a series of instances in which she had "lost her character," or reference, because of moving, presumably from the mill's boarding houses. On one occasion her mother was requested to take her home. Her relations in Lowell were further soured when, after an absence, she returned to find that her overseer, Gaye, had taken away "her" winding frame and given it to another worker. Although her new frame was almost as good, she was furious and believed Gaye too ashamed of his act to speak to her. On her part, she wouldn't speak until he did, "and then I shall give him a piece of my mind for he knows he had no business to give it away." Clearly, she admitted to no hierarchical social separation between her and the overseer, despite his authority. She did not appear to have been learning the lessons intended for her. Instead, she exemplified the continuing difficulties in harnessing the traditionally and potentially independent Yankees to the industrial yoke.[26]

Work rules, as posted and developed by various corporations, defined the owners' ideas of order. They forbade merriment, reading, singing, drinking, meetings, leaving work, or gambling; in other words, activities people had traditionally assumed as part of their daily life and work, for rules are not made against practices not found or expected (no restrictions on ballooning, for example). In all these areas, people had assumed they were competent to determine proper behavior themselves.[27] "Regulations To Be Observed By Persons In The Employment Of The Boott Cotton Mills" dates from the period of the Metcalfs' incumbency. Overseers "are to be punctually in their rooms at the starting of the Mill," reminding them that they, too, are mere workers in the eyes of management. They could grant employees "leaves of absence" only if substitutes could fill their places, and only sickness, with notice, excused an employee from attendance. Workers had to live in a company boardinghouse and "conform to the regulations of the house in which they board." Similarly, "A regular attendance on public worship on the Sabbath is necessary for the preservation of good order. The Company will not employ any person who is habitually absent." The relative positions involved appeared most clearly, however, in terms of the length of employment: "All persons entering into the employment of the Company are considered as engaged to work twelve months, if the Company require their services so long." All the demands were made on the employees by the company, with no obligation on its part to employ for a period, only for the workers to stay while they were useful. Although the regulations were not uniformly enforced, they denote the relationship being constructed.[28]

Employees readily recognized the lack of reciprocity. Mary Metcalf complained, "We have had so much backwater lately I cannot make much." When production dropped, workers absorbed the loss insofar as the company could manage. Later, when the market could absorb goods, she noted that "the speed is on." In neither case did pay compensate for conditions of work; rather, employees worked faster or slower to suit the owners, at consistent, and low, wages.[29]

When work went well, Mary spoke of making "good wages," and hoped to reduce her indebtedness to her family. She remained aware of the difficulty for them of making money on the farm. At other times, she lamented that she had no money to spare. Even as she prospered, comparatively, she passed along the comment that people were "speaking of cutting down the wages." As soon as the production enabled workers to earn those "good wages," ways were being sought to readjust the rates.[30]

In addition to the subjects quoted, the Metcalf letters are filled with discussions of religion, debt, loans, and prospects. They are the product of people trying to find a new place on earth, a new peace with their work, at the same time they worried about achieving the more important peace with their God. The new workers attempted to determine a balance between the independence gained by leaving home and the subservience attendant upon employment by a corporation which wanted to take decision-making out of their hands. They wondered about working for overseers who did not relate to them according to their traditional ideas of fair play and mutuality. They weighed cash wages from the mill against the preferred values of an earlier life and a society where everyone did not simply "look out for no. 1."

What had begun as an experiment had developed into a political and social revolution in management, control, and economy. Gibb compares the new system to an older one: "This was a pattern of feudalism in which the mill was the castle and the agent was the bailiff. Some would have said that, on the one day in the week when he visited the mill, the treasurer was the Lord High Executioner. . . . The position of the unskilled or semi-skilled laborer in Lowell was one of considerable economic and social subservience."[31] The trade from family control on the farm to this subservience did not fit the expectations or plans of many of the women. The Metcalfs and other Yankee women did not tolerate it for long.

Adding financial insult to moral injury, labor was also seen from the outset as the system's economic shock absorber: when business became less profitable as a result of national economic circumstances in the early forties, pay was cut in the mills of Lowell. Marcus Rediker points out the precedent in the background of the owners upon which they may have drawn, noting the practice of "deflecting many of capital's risks in transoceanic commerce on the collective back of the maritime working class" by docking their wages in tough times. In 1843, a committee set up to investigate the Boott's affairs recommended cutting the salaries of the treasurer and agent from $5,000 per year to $3,000. The treasurer, it noted, held the same position "in several others, in one of which he receives the same salary as in this" (John Amory Lowell served as treasurer of the Massachusetts Cotton Mills). While such cuts at top levels did not have a serious impact on costs, they aimed to mollify stockholders temporarily deprived of dividends. More significantly, the committee hoped that the cuts "would tend to reconcile the operatives at the mills to the reduction which has already taken place in their wages and even induce them to submit cheerfully to a further reduction if required." This concept of "cheerful" impoverishment on the part of the workers makes visible the model of expected deference held by the corporate directors.[32]

Dalzell shows how the Boston Associates devised Lowell-style operation in order to ensure themselves and their descendants places of prosperity, safety, and power. They intended to protect their class and to "check the dangerous potential of industry to undermine the peace and tranquil order of society." Investors made every effort to make government a silent partner to their efforts. They "cultivated" politicians, initiated favorable legislation, while steadily augmenting the economic power which they used to foster special treatment of their interests. He makes it clear that their campaign of self-aggrandizement proceeded without regard to its cost to the workers who, because of the emphasis on profit and limited investment (particularly reinvestment), bore steadily increasing pressure to increase production at lower cost to maintain dividends.[33] Expansion of the plant, within limits, also offered the promise of increased profits. The Boott added Mill #5 in 1849, enlarging capacity by about 50 percent and initiating the process which would wall off the city from the countryside beyond.

Immense profits led to reproduction of the pattern, generally with heavy involvement by the same investors, all over New England and beyond. Lawrence, Holyoke, and Chicopee Falls, Massachusetts, Manchester and Nashua, New Hampshire, Biddeford and Lewiston, Maine, spread the new pattern across the countryside. The system worked, and corporate planners came from all over the country to study the original.

National figures came and admired the city and the spectacles presented them. President Andrew Jackson walked through three and a half miles of "girls" the owners had dressed in white, with white parasols and silk stockings, for the occasion. Charles Dickens admired the fresh air of the city, the cleanliness and comfort of the factories, and of the workers remarked, "They were healthy in appearance, many of them remarkably so, and had the manners and deportment of young women; not degraded brutes of burden." They did what they could to improve their surroundings: "In the dressing-room of No. 3 on the Boott Corporation, we counted over 200 pots and plants and flowers." John Greenleaf Whittier described these women as "acres of girlhood, beauty reckoned by the square mile." These descriptions reflected the comparative mildness of conditions in early Lowell and the unusual scale of its operation in the New England countryside, and they carried an implied comparison to the desperate poverty of the employees in Britain's steam-powered urban cotton factories.[34]

The money made in Lowell and its companion cities altered the course not only of local industrial history but also of national development. The Boston Associates' profits, political connections, and diversification led to control of textiles, railroads, insurance companies, banks, and more, giving them a dominant position in the New England economy and significant influence in Pennsylvania, New York, and the Midwest. This closed club of large investors developed substantial political power, serving on the state and national level and supporting chosen candidates such as Daniel Webster, who adopted their views after a gift of stock in the Merrimack Company. Given legislative monopolies in railroads, creating them in textiles, approaching them in other areas, they became multimillionaires. Nathan Appleton alone was involved in thirty-one textile companies, and defender Edward Everett's hyperbole in 1863 described the kind of power he and his cohorts wielded: "Without the aid of William and Nathan Appleton and Abbott Lawrence, no new venture could be launched in New England."[35] Such dominance had previously been unheard of, and it represented a level of individual economic importance which violated the precepts of earlier social beliefs, while heralding the men's success in the new system they were creating.

This system faced continuing opposition from all levels of society. Ralph Waldo Emerson, for one, saw something ominous in the plan to enlarge the Lowell canal system in the mid-1840s:

> An American in this ardent climate gets up early some morning and buys a river; and advertises for 12 or 1500 Irishmen; digs a new channel for it, brings it to his mills, and has a head of 24 feet of water: then carves out within doors a quarter township into streets & building lots, school, tavern,

> & methodist meeting house—sends up an engineer into New Hampshire, to see where his water comes from &, after advising with him sends a trusty man of business to buy of all the farmers such mill privileges as will serve him among their waste hill & pasture lots and comes home with great glee announcing that he is now owner of the great Lake Winnipiseosce, as reservoir for his mills at Midsummer.[36]

Emerson resented the arrogance that permitted itself to purchase natural resources as a source of profit for the few; he suspected the morals of the "trusty man of business" who bought that which seemed worthless to the farmers and the "glee" that resulted from the purchases. The Indian-named lake and river, a communal resource in earlier times, now represented only a millpond and canal for Lowell. Emerson stood for the widespread opposition to the nature of Lowell's development expressed in "intellectual" (read "publishing") circles; he continued his chronicle of disapproval:

> They are an ardent race and are as fully possessed with that hatred of labor, which is the principle of progress in the human race, as any other people. They must & will have the enjoyment without the sweat. So they buy slaves where the women will permit it; where they will not, they make the wind, the tide, the waterfall, the steam, the cloud, the lightning, do the work, by every art & device their cunningest brain can achieve.[37]

Emerson described this "hatred of labor" as if innocently, but what concept could seem more slanderous, more heretical, in puritan New England, where the devil provided work for idle hands? To women he assigned the moral sense (particularly ironic given their subservient position as "mill girls"), and to these men cunning, the sly, underhanded desire for pleasure without effort, the producer of slavery and factories. He offered a biting commentary on the relationship between the two slaveries cited by North and South alike, that of the black, or chattel, slave, and that of the wage-slave selling his or her time to the industrialist without a share in the product. His resentment of the idea that some need not labor, that their triumph was defined as progress, expresses a strong opposition to the beliefs and propaganda of the capitalists.

Workers protested in parallel terms. They renounced a system in which they could only participate by selling their time, in which they gained no stake in the product of their exertion, as they always had previously. They complained that they were made wage-slaves by a system that reduced their role to that of automatons, undivided in treatment or role from the machines they tended.

Strikes protesting such conditions hit textile mills as early as 1820 and continued despite the social disapprobation of such public conduct by women, who recognized the inapplicability of old standards to new situations. Sarah Bagley, one of the workers' leaders and spokespersons, noted that inmates in Massachusetts prisons worked two hours per day less than she and had more time to use their library than did Lowell workers. She observed to an incarcerated forger: "You might have selected some game equally dishonest, that would . . . have made you looked up to as a man of wealth. . . . You might have performed some 'hocus-pocus' means of robbery, without forgery, and passed as an Appleton, a Lawrence, or an Astor in society."[38] Strong condemnation of the nature of the system in which she worked by the "inarticulate" laborer—but recognizing and protesting a situation did not empower her to alter it, only to escape.

Early workers voted with their feet, and they voted early and often, making their way from mill to mill, from town to town, in their search for employment in which they would be better treated. They pursued new situations in which they might profit from their Lowell experience and training.[39]

In the words of Emerson again, "The ways of trade are grown selfish to the borders of theft, and supple to the borders (if not beyond the borders) of fraud."[40] The new order brought a new morality, and in Emerson's view only a new morality would tolerate this order. The depth of labor's opposition appeared in the words of Amelia Sargent, another Lowell worker: "We will soon show these drivelling cotton lords, this mushroom aristocracy of New England, who so arrogantly aspire to lord it over God's heritage, that our rights cannot be trampled upon with impunity."[41] The close parallels between her complaints and Emerson's indicated the widespread nature of the objection to those who made themselves a ruling class to dominate both nature and people in new ways. If Sargent's statement referred to Yankee women, her statement rings true, for they soon left Lowell's mills, only to have Ireland's famines send their replacements. Sargent, Emerson, and many others recognized the extent, the fundamental nature, of the changes accomplished at Lowell. Massachusetts had moved far from its namesake "commonwealth," one of mutual responsibility and effort in which all shared in both the product of their labor and in the deprivation of hard times.

Lowell's first decades witnessed the institution of great changes, the initiation of a modern factory system. The nature of work and many of the bases of human relationships were drastically altered. Yet the new system was also conservative, aiming for limited economic goals and dedicated to providing economic safety and social position for its small group of owners. While work took on new guises, workers remained able to decorate their spaces with plants, confront their overseers without social subservience (if Mary Metcalf is a fair example), and exercise considerable mobility between jobs and across the landscape. The status and skills the women brought to the mills gave them grounds for pride and self-worth in their jobs, for resistance if necessary, for the sorts of protest expressed by Bagley, Sargent, and others.

Chapter Two

A Loss of Lustre, 1850–1870

THE STORY of Lowell between 1823 and 1850 had represented a great new success story for the investors. Visitors to the city expressed their admiration of the neat new factories, their impressive manufacturing capabilities, and their attractive employees. The public relations campaign accompanying the city's early years ran smoothly and with great effect. Lowell had its "Golden Age" in the public eye.

Caroline Ware described the extent of the industry's success as exemplified by altering the consumption patterns of the country:

> By 1860 cotton was indeed king, not only in the fields of the south but in the homes and workshops of the north as well. The annual production of American looms, thirty-six and a quarter yards per inhabitant, was more than twice the estimated average consumption of cotton cloth in Great Britain and three times the amount used by each American in 1830.

Much of the cloth was exported, undermining the comparisons, but the point remains well taken: the flood of cotton cloth had made the industry as powerful in the North as was slavery in the South on the eve of the Civil War.[1]

The flood of material into national and international markets brought new pressures. Costs had to be minimized in order to maintain market share in the face of price competition on staple goods, and control had to be maximized to keep the operations running smoothly. The continuing proliferation of textile mills across the land combined with an unstable economy to increase competition and threaten dividends. Boardinghouses and other "amenities" for workers were de-emphasized and the speed-up and stretch-out became the order of the day. The onslaught of famine in Ireland provided ample supplies of labor lacking even the alternative support of a hardscrabble family farm. International economics made efforts to attract labor superfluous.

THE NEW LABOR

No question remained as to the motivation that brought workers to the mills: economic necessity. Using the Hamilton Company as his example, Dublin describes the changing composition of the workforce between 1836 and 1860. In an expanding operation, the number of native-born females fell from 737 to 324, the proportion of females to males shifted from 85:15 to 70:30, and the percentage of Irish-born women rose rapidly, from 29 percent to 47 percent. Sixty-two percent of the male workers were foreign-born in 1860. The roles of children, the elderly, and males took on a new importance.[2] Dublin also found that by 1860 the proportion of Hamilton operatives in company boardinghouses had dropped to one-third, in contrast to the three-quarters of 1836. Change pervaded the community, the living arrangements, and the mill, where Dublin found the old line of native-born Yankee females predominating only in the Dressing Room (warp preparation). (There the last bastion of Yankee solidarity was retained for a time in the face of a visible new future.)[3]

A vicious circle of events fueled the changes. As the national economy and Lowell working conditions declined, the original workers left the mills and took what strength and organization they had created with them. Without these people, their alternatives, and their beliefs, management could impose harsher conditions on the newcomers lacking options to factory work. As a result, in Ware's words, "the industry's prosperity in these years [1850–60] rested upon the exploitation of labor as that of the earlier period had not done."[4] A difference more in degree than in kind.

HOURS AND WAGES

An 1864 "Time Table of the Lowell Mills" indicated the standard workday for ten months in the year as 6:30 A.M. until 6:30 P.M., with a forty-five minute break for dinner at noon, that time to include leaving one's machine and exiting the factory, arriving at the boardinghouse, and returning, for a few, or simply eating what one had brought, for most. Saturday closing came a little earlier, and the total was "arranged to make the working time for ten years average 11 hours per day." Hours of work had diminished by an hour and fifteen minutes since the mid-1840s.[5]

For the duration of those eleven hours of work, the employees tended more machines than previously, and each machine ran faster than before. In 1867, for example, Boott weavers on a particular fabric received 19 cents per 40-yard piece if they ran six looms, or 20 cents if they ran only five.[6] Compared to the two-loom average of 1836, or the three of Sarah Metcalf's day in the 1840s, either load involved substantially more work, more rush. The rates created great pressure to manage the larger task, for earnings there would be nearly 20 percent greater in steady production. The machinery had not changed substantially, and any improvements played far less a role in the increased load than did a willingness to impart a greater burden on the worker and an inability on the latter's part to avoid it.

Lowell mill wages in general for 1860 offered an insightful perspective on operations, population, and attitudes. In the Card Room, for example, three male jobs paid from $4.48 to $6.51 per week, and female drawing- and speeder-frame tenders made $3.48. Boys, a category new to the Lowell mills, earned $2.70. The overseer of carding received $16.70. Wages ranged from as low as $1.50 for doffer boys and girls in ring-spinning to $21.91 for overseers in dressing and $3,000 per year, or nearly $58 per week, for an agent. Wages of $8.91 for dressers presumably explain the continued predominance of Yankee women there, despite the physical difficulty of the work and the hot, steamy conditions (all of which call into question Dublin's statement that higher wages were not a factor). Overall, his assertion that women's wages were being compressed, their opportunity for wage or career advancement diminished, stands uncontested. The wage summary reflects clearly the standard Lowell practice of hiring the cheapest labor available, along with the artificial hierarchy imposed to restrict to men positions of greater skill and, still more, greater status.[7]

Perhaps more significant as an indication of wages and their meaning to the workers is Robert Layer's analysis of annual earnings, their change and their meaning as a function of the cost of living index for the period. Although the analysis was carried out for the Lowell corporations with records at Baker Library, the generic nature of the companies and their policies in such matters for this period makes the data applicable to the Boott. Layer assigned average annual earnings per hypothetical *full-time* worker of $161.51 in 1825 an arbitrary value of 84, relative to the cost of living. During the next twenty-five years this value fell as low as 77 but gradually rose to 108.2 in 1850. During the 1850s, however, it ranged between 75.9, or less than in 1825, and 95.9, while generally declining. Between 1860 and 1868, the index spent five years below 72, with a peak of 94.8 in the initial year, before rising to about 105 for the new decade. Actual earnings, derived from the periods the factories actually ran, varied still more dramatically, hovering between $160 and $190 per year for most of the time from 1825 to 1863, with a low of $129.72 in 1861, before rising to $327.99 in the inflationary postwar period.[8] The significance of all these numbers lay in the fact that mill workers faced a future of rising and falling wages without real advance and endured a situation which had not improved substantially despite the increase in the amount of work expected, the number of machines tended, the overall decline in conditions within and without the mills, and the rise in production per worker. Furthermore, according to Layer, except for the years 1847–52, cotton textile workers were losing ground relative to all other workers for the periods both before and after the Civil War.[9] This picture conflicts with Scranton's contrasting of irregular but better-paying work of flexible manufacture with monotonous but steady work in bulk processing textiles.[10]

LABOR'S RESPONSE

Despite the declining conditions, only one strike appears to have occurred in the fifties. In 1859 throstle-frame spinners struck the Tremont, Boott, Prescott, and Massachusetts corporations. Three to five hundred immigrant women walked out in demand for higher wages. This small group, perhaps 6 percent of the workforce, could have little effect on operations; their effort had no hope once it became clear it would not attract general support from other workers. Compared to the major strikes of the 1830s involving one-sixth to one-fourth of the labor force, this one revealed the impact of the increased divisions by age, sex, and ethnicity within the mills and the concomitant difficulties in mounting a broad protest. Since these workers struck despite lacking nearly all the sources of support attendant on the earlier strikes, however, the walk-out stood as impressive evidence of the depth of their dissatisfaction. Neither revolutionary-era traditions, concepts of social equality with management, nor the solidarity that came with the homogeneity of a female Yankee labor force that trained new members in the traditions of the old were part of the outlook of these workers.[11] In their place appeared a heterogeneity of workers and a homogeneity of poverty which gave labor a new perspective, one emerging from suffering, mistreatment, overwork, and fear for the next day's bread.

MANAGEMENT ACTION

Management pressure on the mills' production had driven away those who had protested earlier conditions. At this time, the corporations combined in a variety of official and unofficial ways to enhance their advantage. In general, "the ability to present a solid front to the laborers is their primary purpose."[12] Through various organizations they regulated wages, output, politics, and other matters. Suffolk Mills treasurer Henry B. Call relayed to his agent, John Wright, the directors' decision in 1850 to cut production by one half on account of a stockpile of cloth, the high price of cotton, and uncertainty regarding improvement in demand. The presence of this letter in the Boott files indicates the cooperative nature of factory management in Lowell.[13] Close ownership and a belief in shared values, needs, and purposes overwhelmed any thoughts of competitive secrecy or individual advantage. In fact, through the New England Cotton Manufacturers Association (NECMA), they exchanged information on technological advances and compared assessments of mill operations. James B. Francis, widely recognized as the dominant force in the powerful Locks and Canals company, engineer of the landmark Northern Canal, creator of the millpond Lake Winnipesaukee, and waterpower innovator extraordinaire, revealed another aspect of their joint actions in a letter to the Boott treasurer in 1860. In it he assured T. Jefferson Coolidge of the availability of a loan from the City Institution of Savings for $20,000–25,000. "In the ordinary course of events they expect to receive nearly double that sum by the close of the January quarter. . . . Most of the money in our savings banks comes from the manufacturing companies in the first instance, and I think there is a disposition in our institution to let them have the benefit of it as far as they can, in such times as the present."[14] Thus the corporations controlled the supply of money for continued investment, freeing their "own" money for dividends.

A letter from F. B. Crowninshield, treasurer of the Merrimack, to Coolidge demonstrated a more formal and pointed cooperation:

> Several of the Manufacturing Companies at Lowell having expended sums of money by the payment of counsel fees, printing, etc. for the purpose of effecting a repeal of the shares law, last winter, will you please send me the amount you have paid, if any, that I may apportion the same for settlement, according to the understanding and agreement of the united corporations.[15]

Joint political action and prearrangement for shared expenses ensured maximum effort and minimum risk for all concerned. The corporations were assured a significant voice in the "public" forums of the State House.

The various companies also formed a mutual insurance company to work to prevent the outbreak of fire and to spread the risk of losses to fire. Through such companies as the Boston Manufacturers' Mutual Insurance Company they circulated careful accounts of larger conflagrations in order that all might learn from the accidents of one.[16]

GOVERNMENT ACTION

Given the power inherent in the mills' scale, coupled with that garnered through joint action, it comes as no surprise that they operated with virtual freedom from interference by employees or the state. The commonwealth's concern for its citizens first appeared, in this connection, in the creation of a state board to oversee local relief charities as early as 1853.[17] In this way the state expressed a willingness to superintend and a feeling of need for oversight of those for whom the towns were expected to care. That action suggested an initial perception that the new system did not attend to those not serving it.

Massachusetts also moved to protect government interests and passed the "shares law" over the objections of the corporations in 1863. The new law required the state to withhold 15 percent of every dividend paid to out-of-state stockholders; in-state investors were taxed 3 percent.[18] Despite the power of the corporations and their joint efforts, they were unable to prevent the state from seeking to share in the profits of the new industrial system.

MANAGEMENT CRITIQUED

Boott management came under heavy fire from stockholders as well as workers during the fifties and sixties. One of the most vociferous critics was Boott stockholder James Cook Ayer.[19] Ayer had made his fortune in the patent medicine business in Lowell and then invested substantially in a number of textile mills. He stood outside the original circle of investors on all counts: with regard to geography, family, training, and other shared ties. Not part of the Boston club, he objected publicly when he felt corporate performance not to be in the stockholder's interests.

In 1863 Ayer, in *Some of the Uses and Abuses in the Management of Our Manufacturing Corporations,* assailed corporations' management. He complained that stockholders' meetings were held in tiny Boston offices, called on short notice and/or simultaneously for different companies, papered with false proxies, and held before the arrival of the Lowell train in the morning; directors withheld addresses of stockholders even after the state required them to be available. Corruption was a way of life (as Emerson had feared), as instanced by an $89,000 "bribe for Congress" noted, without comment, at a Middlesex Company meeting.

Ayer found the benefits of the industry distributed among none but selling agents and managers controlled by and beholden to a few of the investors, and that "while these grow rich, the [other] owners, skilled workmen, and operatives are poor." With no check on these officers, their power expanded while their efforts for the companies diminished, he felt. When the legislature did pass a law in 1859 requiring mild reform, the companies ignored it until able to secure its repeal, partly through petitions coerced from the workers by threats. Ayer characterized the situation as one of "wanton robberies" of the stockholders. He cited as an example two selling agents, A. & A. Lawrence and Company (the Boott's agent) and J. W. Paige, who received $200,000 apiece annually in commissions and as much again in incidentals, without capital, risk, or expense. Boott stockholders' efforts to reduce payments to Lawrence & Company had been foiled by the presence of a principal of that firm on the Boott's board of directors, he charged. Abbot Lawrence, a director, had positioned himself to receive large profits from his roles in both the selling houses and the corporation. Large commissions for purchasers of cotton, coal, wood, and oil appeared to Ayer similarly exemplary of the nepotistic robberies taking place. Extremely high salaries for the officers went to "men who have failed to be very valuable in any other pursuit—the son, son-in-law, nephew, or relative of some Director, who, in turn, allows the other Directors to put their dependents in good positions, also." The result for the workers in the mill was evident: "The labor employed in them has been pressed from low, even lower, and down to the least per diem that will support life." Ayer condemned not the individual managers/directors, but found "that the *system* is thoroughly bad." Reform could come only through state regulation: "So vicious had this system become in its influences, that it might almost seem to make swindlers of virtuous men, and it should be changed, not alone for the preservation of the great properties it concerns, but also to save from contamination those who administer it."[20] While echoing Emerson's sentiments, Ayer defended those involved in all levels of the operations except for those at the very top, primarily in the Boston offices. He portrayed the industry as bloated by success, monopoly, and a willingness to operate for the benefit of its smallest component.

Here is Ayer's description of the appointment of Boott Treasurer Coolidge in 1858:

> The stockholders of the Boott Company, finding their property depressed by mismanagement, desired to select some able and practical man for a Treasurer, who could restore the Corporation to a prosperous condition. Attempting to move for this purpose, they were met with the assertion that the Company was largely in debt; that one of the Directors, a wealthy man,

> would furnish it with money and with credit, if his son-in-law were elected Treasurer, and without such aid the Corporation must fail. It was not claimed that their candidate had any special qualifications of fitness for the place, except having married one of the Director's daughters.[21]

Boott director Nathan Appleton was Coolidge's father-in-law. After his appointment, in 1859, the directors of the Boott sought and received permission to issue $300,000 worth of stock authorized but not issued in 1837.[22]

THE TREASURER COOLIDGE ERA

T. Jefferson Coolidge, treasurer from 1858 to 1865, arrived at the Boott to find it in poor financial condition, and he left it the same way. Interestingly, this fact had negligible impact on his career. In 1868 he became treasurer of the Lawrence Company and the Amoskeag Company (of Manchester, New Hampshire), and he ultimately held directorships in over forty companies, as well as seeing public service on the state and national levels.[23]

Coolidge's actions and reports illuminated the Boott's attitude during the Civil War. In 1860 he indicated the expenditure of $5,394 for speeders and stretchers for the No. 2 mill, a pitifully small amount to spend on machinery for a year's time.[24] His accounts of the corporation's status and operation during two consecutive six-month periods ending at the close of April 1861 revealed much about the mill's history, situation, and plans at this pivotal moment.[25] At this time, after the "depression years" of the 1850s, the corporations had begun "to pay closer attention to cost accounting"; in fact, according to Chandler, "it was not until the 1850s that the owners and managers began to use their accounts to determine unit costs."[26] The six-month accounting offered a detailed example of this new attempt at precision in such matters. Coolidge reported that the "fine Mills" had produced large quantities of satisfactory goods, but that while "coarse Mill No. 1" had done well, "the product of Mill No. 4 has been 47.885 pounds, 141.66 yards short of the same season last year, and the cost of labor has been very high, 3.79c. This is owing partly to the changes made in the machinery." (It needs to be noted that the terms "coarse" and "fine" represent comparative judgments related only to this corporation. The former utilized #14 yarn, the latter #30s, both in the coarse category according to trade standards). The separation indicated for the operation of the different mills negated some of the economies of scale associated with the Lowell mills and added to the complexity of the treasurer's and, especially, the agent's task. Chandler asserts, "As the mills were designed to facilitate the coordination of flow through the process of production, the mill agent's administrative task was relatively routine." In this scheme, however, cotton would be purchased for two grades of cloth, and then the flow of fiber, rovings, and yarn for the various materials produced would have to be managed separately for each of the four mills, an approach that demonstrated little of the expected simplification associated with a large integrated unit. Changes in product, coordination of the flow, and oversight of purchasing and of administering the workforce presented ample challenges, and these were multiplied by the fact that the complex contained increasing numbers of mills, each of which operated as a nearly autonomous production unit.[27] The Boott ran spasmodically, responding to a number of independent factors, including availability of water for power, with the national economy and the state of the market for both cotton and cloth further complicating the flow.

Coolidge's report continued by describing the limiting factor with regard to the coarse product: "The Drillings made in both Mills, though very heavy was not as good as could be wished, nor do I believe that with the present buildings and Machinery, they can be manufactured to compete with the Pepperell or any other of the best marks." Thus mills nos. 1 and 4 appeared to have problems that resulted from the conduct of previous treasurers. For the operation to have reached this state of affairs revealed either a lack of knowledge over the past decades, or knowledge without willingness to remedy developing problems, possibly at some expense to dividends. Given Coolidge's brief tenure, I assume his information came from the agent whose complaints were aimed at matters beyond his own responsibility and before Coolidge's appointment.

Coolidge could point to few changes for the better during his years in charge. He had dismissed one overseer and made one man responsible for spinning in both coarse mills (physically separated by the two fine mills). He had also made a change aimed at addressing the issue of employee loyalty: "The examining and classing the goods has been removed (at my suggestion) from the weaving rooms where it was for the interest of everybody to make as few rejections as possible, to the Cloth Room, and though the number of seconds has been much increased, the quality of the firsts has been proportionally raised." Significantly, after some twenty-five years of operation, the mill's management still sought to create a structure within which the employees, at all levels, would be ruled by management's desires rather than by what the workers saw as their divergent interests. Despite the fact that cloth coming from the looms was evaluated, or classed, as to its quality, and despite the fact that workers involved in the task, and their supervisors, were employed to identify faults and to assign blame for them (e.g., to carding, spinning, loom malfunction, or weaver error), according to which determination fines and loss of wages would be assigned, as long as the operation retained a single characteristic of joint effort (the fact that classing and weaving took place in the same room), it encountered labor solidarity blocking management-directed assessments of quality.

Similarly, a letter of recommendation sent to Coolidge for a candidate for agent concentrated on moral issues rather than knowledge. Based on twenty years of acquaintance, the writer could vouch for the candidate's trustworthiness and honesty, his energy and willingness to "devote himself soul and body to the interests of your corporation." Regarding his ability, the writer could only say that, "As far as I can learn, [he has] sufficient skill and knowledge of manufacturing."[28] In the writer's view, the character traits that produced loyalty to the corporation rather than to self marked the attractive candidate, a position which implied a scarcity of such people, a lack of complete success in the reworking of human nature to fulfill the demands of the new order at even the highest employee level. The continuing difficulty of creating a situation in which workers were led to identify with, adopt, or even serve the owners' interests rather than those of their cohorts created another level of operational difficulty heightening the contrast with the common image of smoothly running giant operation associated with the Lowell system.

Coolidge finally turned to the larger issues of the corporation and asked the directors to consider, "if the China War should continue" and cause drillings to accumulate, if they would choose to enlarge and repair Mill No. 4 and replace whatever machinery there was defective. Otherwise, he said, "I do not think it would be worth while to do anything, unless so thorough a change were made, as would enable us to compete in cost of Labor, regularity of Picks, & eveness of yarn, with the Jackson, or any other of the newest and best Mills, and this would be a matter of very great expense, requiring deliberation." The treasurer felt he was managing a mill that could not compete with others in Maine and New Hampshire with newer equipment and buildings. The strategy of postponing maintenance, let alone improvements, while maintaining dividends had reached its predictable limit.[29]

Coolidge's list of areas to be improved indicated the extent of his perceived disadvantage relative to other mills. A letter to him from S. J. Wethrill of the Otis Company, in Ware, Massachusetts, described the problems with piecemeal improvements. Regarding carding in the No. 4 mill, he noted the lack of space and the necessity of moving other machinery and the shafting system. Only if new cards were sufficiently improved to eliminate double carding, thus halving the space needed, would a change be practical.[30] Coolidge chose not to trust in the claims for the new cards and did not proceed. The fact that he had sought an outside opinion on such a routine matter did not speak well for either his expertise or his confidence in relying on his staff.[31]

In the six months ending October 22, 1860, the Boott had produced 3,746,164.5 yards of drilling and 3,876,784 yards of fine goods for a total of 7.6 million yards (before finishing). Cloth sold netted $528,844.35. Costs and profits were calculated to four decimal points (32.5024 cents per pound grossed for fine goods), indicating the effort at careful bookkeeping Chandler noted. The profit from manufactures equaled $97,046.65, nearly 20 percent of sales for six months in a comparatively poor year, on reported investment of $1.2 million. Rents and other income swelled the figure to $104,194.79, after deducting interest of nearly $11,000. Addressing the path by which such a profit had been obtained, Coolidge explained: "The reason of a profit having been made this 6 mos. as large as the preceding six, although the sales were much smaller, and the Goods brought about the same prices" are a drop in the price of cotton, a saving in interest payments, and "by an error at the Mill, some thousands of pounds of Roving [were] omitted in the last accounts." The on-going relationship between the mill and its power supplier, Locks and Canals, also improved appearances: "During the past 6 mos. a Dividend of $10 on our Stock in the Locks and Canals [equals] $5,360.00 and have not been required to pay the usual annual assessment, $300 a power [equaling] $5,360.00." He described a situation in which dividends from and payments to the power company canceled out, providing free water for power to the corporation.

Finally, Coolidge placed the last six months in a broader context reflecting the Boott's overall performance during 1858–60:

The earnings of the Company the past 2 1/2 years have been without taking out Repairs, and a small balance to make up Deficiency of Capital	$586,309.88
or nearly 50% [on investment] of which there has been spent in Repairs and Renewals of Machinery	$121,720.25
and on Dividend	$ 48,000.00
leaving a Quick Capital of	$416,589.63
which will be diminished by any Dividend you may see fit to make.[32]	

It appears that the mill had been doing rather well for the close of what Chandler describes as "depression years," with earnings of nearly 50 percent of the owners' investment in two and a half years. Coolidge's statement revealed that a large percentage of the company's earnings moved into "Quick Capital," a pool of liquid assets over which the directors exerted control and from which they could draw as they saw fit. One effect of this arrangement was to make the mills worth more in assets than suggested by the nominal value of their shares; in Melvin T. Copeland's words, they were "undercapitalized," making their dividends appear low, since they were actually "paid on an investment larger by the amount of earnings turned into improvements which are reflected in the high [selling] price of shares."[33] In the Boott's case, the money appeared historically, and at this time, to have been paid out or held separately as quick capital more than used in repairs, thus explaining the poor state of the facility. Of the amount above assigned to "Repair and Renewals," a significant portion presumably went to the installation of a Warren waterwheel to power the Picker House. Unfortunately, Francis reported to Linus Child, the Boott agent, the wheel was "not quite of sufficient power to carry all the machinery at the usual rate of speed," largely because of an error in planning for it which provided insufficient depth in the tailrace for "free discharge." While remediable, it does not indicate technical competence at the mill.[34]

Six months later, Coolidge reported an increase in the amount of goods made but not sold, an approximately fivefold increase in bad debts (to $20,406.36), and an improvement "in our condition of $29,450.45 or not quite 2 1/2 percent." Coolidge did not blame the war for the problems; inadequate, outmoded machinery was the cause: "Of 896 looms on 30s [fine goods] we have unfortunately 448 on narrow shirtings & print cloths not over 30 inches wide, & 250 more on goods of only 35 inches." Approximately 700 out of 900 looms for the "fine goods" were incapable of making cloth in salable widths. The looms' condition reflected long-term mill practice in continuing to operate equipment after it outlived its usefulness. The situation was the more unfortunate since profits on fine goods, as reported six months earlier, were nearly 25 percent of production costs, much better than on the drillings. As these archaic looms were being shut down to avoid losses, workers were rapidly being laid off. The war played no role in these developments, indicating that reasons other than the price of cotton were directing events at Lowell.

Little other information survives to describe Boott production or sales for this period. Evidence indicates that the Boott product, for reasons of quality or efficiency, could not be sold at a price sufficient to repay them for the cotton and their trouble.[35] Still, when the Boott decided to drop the A. & A. Lawrence and Company as their selling agent in 1865, they received inquiries from at least seven firms, from which they chose George C. Richardson and Company of Boston. By September of that year they had $275,660.56 worth of goods committed to Richardson.[36]

As a response to the desperateness of the mills' condition, Coolidge wanted to stop. The machinery in the fine mills was no longer useful, and the condition of the coarse mills made the Boott unable to compete. Since production was therefore not "desirable," Coolidge wanted to overhaul at least the poorest mill, No. 4: "The Mills are old, have been run very hard, & though every 6 mos. larger sums comparatively are spent in repairing them than I believe in any Mill at Lowell, the goods produced cost high, the waste is too large, the yarn uneven, the cloth not what it should be." With $387,320.08 in Quick Capital and a "permanent loan which is not likely to be called in" of $315,000, Coolidge felt the time was ripe for rehabilitation of the physical assets and looked for the appointment of a committee to consider his suggestion. He described an operation that had been run into the ground, saddled with outmoded machinery and deteriorated buildings, to the point where it could no longer compete.

Many of the Lowell mills closed to various extents during the Civil War. There were profits to be made selling their large stocks of cotton at prices inflated by the secession of the South, buildings and machinery to refurbish, and for some, at least, expectations that the war would end quickly, permitting easy replenishment of supplies of cotton. Perhaps their sympathy for their erstwhile and long-defended trading partners in the South, the plantation owners who supplied the cotton and bought cloth, misled them regarding the extent of the national division they had worked so long to deny or prevent. For most of the workers, the stoppages meant substantial periods out of work. The contemporary historian Charles Cowley condemned the "ruthless dismissal" by which the nine corporations "in cold blood" sent 10,000 "operatives penniless into the streets."[37] A recently discovered 1861 letter from General George Stark, head of the Boston & Lowell and Nashua & Lowell Railroads, to Coolidge suggested worker objections were immediate and strong:

> The undersigned has reason to believe that threats are made by persons in Lowell, to burn any cotton which may be sold to remove from that place for manufacture elsewhere.
>
> He therefore deems it his duty to hereby notify you that the time and manner in which such cotton is now being delivered to the Railroad for transportation renders it peculiarly open to incendiary attempts.

Coolidge disputed the validity of Stark's denial of responsibility for any damage, but not the information in the letter.[38]

Coolidge polled the directors in 1862: "The Boott Cotton Mills are stopped for repairs which will take three months. They have in hand about 1300 bales of cotton at a valuation of 12 cts., the market value being 50 to 60,000 dollars above the valuation." He went on to ask if he should sell all or any part of the cotton. Two directors indicated they wished him to sell one half, one, F. B. Crowninshield, serving also as treasurer of the Merrimack at this time, supported selling all, another "a considerable portion." J. A. Lowell, a previous treasurer, offered a thoughtful view:

> I do not, as you well know, think that is ever the true interest of a mill to stop for the sake of a profit in the cotton.
>
> Your case is a little different and the question as I view it is this: Is it expedient to sell cotton now in the expectation of buying it cheaper three months hence.
>
> I do not think it is.[39]

None of the replies alluded to the issue of the workers' interests, and only Lowell indicated he felt that the mill should make its income from manufacturing rather than speculation in cotton. The other directors appeared to follow the "modern" concept of maximizing income over the short term and determining later how to continue the venture. Interestingly, it was not the mill manager, as Steven Lubar suggests, who mandated the sale of cotton, but the directors.[40]

Ayer found their program failed even as speculation:

> When the rebellion broke out, it [Boott] had its usual quantity, about 2,000,000 pounds, of cotton on hand. The advance on that alone, if it had been retained until now, instead of being sold, as a part of it has been, to go abroad, would have been sufficient to pay the whole dividends of the last fifteen years, the repairs [to buildings] above mentioned, and left a surplus on hand.[41]

According to this evaluation, not even short-term interests of the investors were served.

The Civil War does not appear to have halted the supply of cotton to Lowell or to have required the mills' shutdown. Other centers, such as Philadelphia, remained in operation, and cotton continued to be available in world markets.[42] Problems did exist and some, at least, were predicted. As early as December 1861, the Boott received warnings of shortages in specie caused by British efforts to undermine the North's economy. Lack of a medium of exchange remained a problem for some time and added to the difficulties of obtaining cotton or loans.[43] As J. A. Lowell had expected, cotton prices rose steadily, from the low teens at the start of the war to around 55 cents by the end of 1862, when steady supplies were reaching Cairo, Illinois, "owing to the possession by the Federal army of the rich cotton growing country around Holly Springs,

Miss." Rebel attacks on railroad lines caused inconvenience, as did a railroad strike in Providence. Although cotton remained available, its price climbed above 80 cents.[44] Early in 1865 George Dexter, who had been buying for the Boott, wrote to ask what to do with their gold "if you are going to shut up the mill and buy no more cotton?"[45] Despite repairs already made, despite the availability of cotton to the Boott under the same circumstances enabling others to profit during the war, despite the thousands thrown out of work by the mill's inactivity, he found no resolve to continue manufacturing. Instead, the treasurers of the Lowell corporations signed an agreement at the end of 1864 altering their mutual insurance to permit any of them to manufacture wool, as if a change to this more difficult, higher-skill production would have provided a solution to their inability or unwillingness to operate.[46] Typically, two treasurers were able to sign for a total of five companies.

The Boott used this time to construct two new buildings, a Picker House in 1860–61, and a Cotton Storehouse in 1865, the latter served directly by a railroad line running through the once-pastoral millyard. Between 1863 and 1865, they connected mills #1 and #2, and #3 and #4, and spent $212,000 on new machinery for two mills. Towers exhibiting a bell and a seven-foot, four-sided clock were erected at the junctures. This construction completed the chain of development, begun with the construction of #5 in 1849, which walled off the city from a view of the countryside. Such construction need not have interfered with production significantly. Furthermore, extensive replacement of machinery took place in only two of the five mills, leaving the lion's share of the capacity continuously available. Neither lack of cotton nor extensive rebuilding would have required extended closings, therefore, as has often been assumed.[47]

The operation for which it had been claimed that it ran as a "public good" was no longer a novelty. Instead it stood amidst an array of competitive forces. It had become clear that despite the fact that Lowell was a pretty city, its factories did not operate so differently than England's "satanic mills." Dalzell describes the Boston Associates' desire to create an economy that would enrich, support, and protect them. In fact, they bent all efforts, political, economic, and social, to restrict and contain the economy. They resisted continuing expansion even when it could have increased dividends, instead preferring to keep costs down and the operation controllable, to reserve the benefits of the new economy to their own group. Even growth of the system was resisted once it had become sufficient to this purpose, despite the fact that it would have increased employment and dividends.[48] Instead, steadily increasing pressure was brought to bear on labor in order to maintain profits. The

Boott's dilapidated condition and its antiquated machinery were exemplary of these developments. Extreme unemployment during the Civil War was an avoidable result, brought on by the perceived opportunity for profit without manufacturing. The Boott Cotton Mills, an operation in trouble, entered and left the Civil War period bedeviled by old buildings, antiquated machinery, and a management devoted to short-term profits.

Part Two

Expansion and Failure, 1871–1904

Chapter Three

Organization and Equipment

At this time, the point where histories of the Lowell mills have generally ended, material for discussion of the Boott Cotton Mills (BCM) begins to increase dramatically. The quantity of data available facilitates our description of the corporation's operation, its management, technology, and labor, as well as their relationships with one another. For example, the equipping of a new mill, #6, at the start of this period revealed aspects of the thought and intentions of the company and established patterns of operation which would affect the mill for decades. (Buildings were traditionally designated either "no. 6" or "#6," with floors: e.g., #6-1, #5-2.) Clearly, the complement of machinery was not static, and its changes over time impacted the workforce and the products. On the other hand, a mill is much more than a collection of machinery. It has a population of its own, workers and managers, whose daily lives are intertwined both with one another and with the equipment.

The complex grew and closed off the city from the view of the river and beyond, creating an interior courtyard featuring ordered plantings and walkways. Later in the period, it became a sparer scene, with plantings and most structures removed. Two sets of railroad tracks entered the yard to bring coal and cotton into the complex and take cloth from it. Only by crossing the bridge adjacent to the Counting House could workers move in and out of the mills and horses bring supplies to the loading docks at various buildings. Access to the mills was thus closely limited, and the complex was separated from the community by architecture as well as by ownership.

Overview and Trends

Having largely ignored cotton manufacture for the sake of windfall profits on cotton fiber during the Civil War, by 1870 directors saw the potential for increased manufacturing profit through expansion. The

Lowell model of large production runs of a limited number of styles was tried and true, and the market was rebounding.

Most purchasers of the mill's cloth were hidden by the intermediary of the sales agents, now Smith, Hogg, and Gardner of New York, but the Marshall Field Company was one of the buyers. Shipments to "Hayti," other West Indian destinations, and Spanish Honduras indicated significant markets, as did sales of warp yarns, on beams, to the Pacific Mills in Lawrence and the Hamilton, in Lowell. The treasurer explored a Korean market for the drillings and sheetings. New products, new markets, and new problems, as we will see.[1]

Despite the ups and downs of cotton manufacturing, enlargement of the physical plant continued fairly steadily during this period. During 1871–72, the Boott added the six-story Mill #6; shortly thereafter, one story was added to the Picker House, and two towers and a fourth story were added to Mill #5. Ten years later a new cotton storehouse replaced one of the boardinghouses, Mill #5 was extended to the east, other buildings were connected to one another, Mills #1–#4 gained a fifth story, and #7 rose two stories.[2]

Mill #6 enlarged the plant's capacity in carding, spinning, and weaving. It also contained a blacksmith shop, a large machine shop, a paint shop, and a carpenter shop.[3] In most respects its operation was self-sufficient: it operated as a separate mill, as had its predecessors. The new mill added to the Boott's production of sheetings, shirtings, and drillings. It raised the operation's statistics to 112,752 spindles and 1,875 workers processing 6,760,000 lbs of strict middlings cotton into 23,920,000 yards of cloth per year in 1876, more than a 50-percent increase since 1870.[4]

As part of the plan to produce more cloth, the owners enlarged their power supply. Steam power had been introduced in the 1860s, and in 1873 they advertised the addition of a 440-horsepower (HP) engine. By 1878 they operated a 1,000 HP Corliss. Steam increased the total power available and insured against slowdowns when water was scarce and business plentiful. Succeeding engines installed in the 1880s and 1890s made steam and waterpower approximately equal partners in powering the mill complex.[5]

Waterpower at the Boott in 1870 consisted of six breast wheels of some 100 HP apiece, the Warren wheel powering the Picker House, and the two Francis turbines of about 230 HP, for a total of over 1,000 HP, except when high water in the river below the mills caused backwater, partially submerging the wheels and drastically limiting their effectiveness.[6] By the time of the building of #6, however, a new generation of turbines had become available: "Early in the seventies the Swain wheel, one of the first successful inward discharge turbines, had been thoroughly tested by Mr. Hiram F. Mills and Mr. Francis, and was offered to manufacturers as one which not only developed great power, but was susceptible of being installed above the tailwater and moderate backwater."[7]

Since the turbines did not need to be as large as the amount of head available, as the wheels had, they lost only the percentage of power represented by the extent to which backwater diminished the available fall of water.* During the seventies and eighties the Boott replaced all its wheels and turbines with eight 72" Swains generating 300 HP each and one 80" Swain of 400 HP, for a total of 2,800 water-generated HP. Most of the difference between this figure and the power available previously came from the increased efficiency of the new wheels. While the breast wheels had been 40–50 percent efficient, 60 percent at best, the Swains offered up to 86 percent efficiency. Where possible, workers installed the turbines in existing wheelpits. Stone drilling, stone cutting, and bricklaying adapted old breast-wheelpits to hold up to three of the new turbines.[8]

Lowell investors had long been committed to the idea that limited development provided sufficient income and served their conservative interests. Now, "very simply, as an experiment in developing a distinctive kind of industrialization—one that combined modern practice with the familiar patterns of traditional society—the Waltham-Lowell system was dead, the victim of a process of change the relentless logic of which the Associates chose, finally, not to resist."[9] The way to increase profits was to enlarge the scale of operations, and the Boott's performance in the final quarter of the century exemplified this belief:

	no. of spindles	yds of cloth/week
1870	84,000	350,000
1880	113,000	...
1890	151,000	830,000

Similarly, production rose from 350,000 yards of cloth per week in 1870 to 830,000 per week (43.1 million yards per year) in 1890. Increases flowed from a combination of sources. More spindles, higher spindle and loom speeds, and lighter cloth all contributed to the increasing yardage.

*Seventeen feet of head required a wheel of at least seventeen feet to capture it's power. When the river was high, backwater retarded the wheel and curbed its power. A turbine did not need to equal the head in height, and thus lost only the percentage of power represented by the rise in the river below the mill: a one-foot rise cost it one-seventeenth of the available power. A wheel turning through that one foot of water in its wheelpit, however, lost substantially more.

Through the 1880s, production was limited to a few kinds of cloth: sheetings and shirtings represented about three-quarters of the total yardage, but drills, twills, and sateens figured increasingly prominently, as did plain cloth made for printing or conversion; coarser ducks played a smaller part than before. Average weight of the yarns produced ranged from 20s to 38s. By 1889, however, fifteen different materials were being produced, according to drawing-in records, and by 1895 a number of other fabrics were added.[10] Production records from the nineties show the addition of diaper cloth, linings, dress goods, and linens to the mill's line, as well as experimentation with corduroy.[11] By 1902, the extent of the change in tactics was still more obvious: "Linen towelling, Homespuns, Cotton Dice Crash, Duck, Jute, Diaper Cloth, Argyles, Drills, Bedford Cords, Oxfords, Corduroys, and plain cloths; all in considerable variety of weave and weight, and with some colored yarns."[12] Variety increased, runs shortened, and cloth was made to meet orders, not to swell inventory. These changes in product had a profound impact on every aspect of the Boott's operation.

By the end of this period, warnings sounded regarding the mill's products and policies. In 1902, William Parker, a widely respected consultant from Lawrence, Massachusetts, delivered a survey of operations that charged, "It would appear from a study of the kinds of yarn used and the character of the cloth manufactured, that the mill has attempted to make too great a variety of yarn and cloth." The plant that had begun and for years had run on three types of cloth made with a narrow range of coarse cotton yarns now spun, in a six-month period, warp yarns of many different weights, or counts (as in hanks per pound), numbered 7, 8, 10, 13, 16, 18, 20, 22, 28, 40, and 50; weft spun on ring-frames: #11, 17, 20, 22, 24, 30, 50, 60, 70, 75, 80, 85; and "of Mule Filling, every number from 7 to 85. Here are over 100 different kinds of yarn made of several varieties of cotton, linen, flax and [cotton] waste, to be watched and kept separate through every process of manufacture. During the same time 276 different styles or kinds of cloth were made."[13] This variety and fineness represented a new path for a Lowell mill, one fraught with perils for a mill where managers, employees, and machinery were adapted to a simpler production. It matched the variety of the Philadelphia manufacturers that Philip Scranton describes who had always specialized in this style of operation, employing plants, equipment, and labor intended for flexibility.[14]

Parker also identified a cause of this changing production: Southern competition in coarse goods. However, despite the South's lower costs, he claimed that Lowell "is still a good place for a good mill, and I believe it is possible in the long run to succeed there about as well as anywhere else." He then initiated a complaint which would continue for decades: "I cannot see how it is possible to make a satisfactory rearrangement of the machinery in the old Boott Mills, and would not recommend the expenditure of a single dollar in that direction." He noted a "lack of concentration of machinery belonging to different processes," which found cards in seven rooms, roving frames in eight, ring-frames in seven, mules in five, spooling in two, and looms in eight, "with some of the rooms so small that the expense of caring for [overseeing] them is more than double what it should be."

When it came to suggesting a solution, the consultant offered bluntly:

> Your old buildings have perhaps served well their purpose in the past, but they were long ago out of date, and are of no value now even if they can be considered safe to work in.
>
> I therefore recommend the entire demolition of the present structures, or at least so much of them as are dangerous to work in, or would in any way interfere with the best arrangement and construction of a first-class new mill.[15]

Buildings from before and after the war failed to meet modern needs. Recent construction had not been designed to meet contemporary demands.

Overall, the period was marked by major changes in the manner of operation, from long runs of coarse cloth to varied production of finer goods to varieties of cloth made to order rather than for inventory. The extent of these changes made the Boott a different operation than in the past.[16]

THE AGENT

Increased rationalization of operations in the postwar period restructured management, thus offering an answer to James C. Ayer's complaints on that score. The Boott's agent during this period was Alexander G. Cumnock, who held the post from 1868 to 1898. Cumnock came to this country from Scotland in 1846 at the age of twelve, immigrating with his parents, who farmed in Massachusetts and New Hampshire. A.G., as he was known, attended Lowell schools and spent two years at colleges in Lowell and Boston before entering the Hamilton Manufacturing Company, then moving on to the Boott. An apt student, he moved away to become the agent for the Quinebaug Manufacturing Company in Connecticut before returning to the Boott in that position two years later, in 1868.[17] Thus he followed the path that would become standard in the late nineteenth century, when it finally became common for a man to attain such a job by demonstrating his expertise in the operation of a cotton mill or a machine shop away from Lowell.[18]

Surviving correspondence illuminates the roles played by treasurer and agent and their relationship to one another. When Treasurer Augustus Lowell went to Europe for over two months in 1882, he simply left his son to act in his stead. Treasurer E. C. Clarke's visit to the mill in 1890 was treated as a special occasion.[19] Neither treasurer appeared to have had much to do with the actual operation of the mill, acting instead as a conduit to Lowell of money, coal, cotton, and other major supplies purchased under Cumnock's direction or for his approval.

The task of keeping the mill supplied with cotton was made difficult by the weather, with its influence on the availability and cost of cotton, as well as by past and future growing conditions, foreign supply and demand, acreage planted, and interest rates, present and future. Brokers provided Boott treasurers with advice and filled their orders, based on the agent's specifications.

Once ordered, the fiber became the agent's concern. He reported enough "lost in transit" to operate a sizable mill, noted bales "water-packed and rotten," others "low grade and dirty," short deliveries, or other failures to meet expectations or standards. The treasurer also ordered coal, again to the agent's specifications as to type and amount. In other areas of major expenditure, the purchases were made through the Boston office but entirely at the agent's behest. When an order for timbers, for example, did not indicate whether or not they were to be planed, the treasurer had to write and ask.[20] Communication of purchases needed to be detailed but not discussed. Relations between treasurer and agent do not seem to have been strained. When an order of 7/16" rope arrived late and short, Cumnock wrote: "Probably the rope was disquieted at the slow time made between Boston and Lowell . . . and shrunk 10 lb."[21]

On the scene in Lowell, the agent continued to have a great deal of authority over operations and employees. Correspondence reveals he hired down to the level of second hand (in charge of the second shift). Personal recommendations, advertisements, and inquiries from interested parties all led to employment. The Counting House, of course, was his particular domain, and there he hired and promoted. He communicated with the various overseers by note, at least until 1890, when he ordered the "new metalic [sic] system" from New England Telephone and Telegraph.[22] Production reports and cost analyses enabled the office to follow operations, and overseers' time books and payrolls enabled it to track, check, or correct wages. Cumnock revealed a susceptibility to gadgets in these areas. In an unsung example of Lowell providing leadership and offering advice to the industry, he wrote to the Vassalboro [Maine] Woolen Mills that to figure payrolls he used "Brisbane's Golden Ready-reckoner," as well as that of Philetus Burnham, paymaster of the neighboring Prescott Mills. Similarly, he sent a testimonial to the maker of a letter-opening machine: "It seems as though every one with much correspondence must have one of your machines in order to keep up with the times, as it works like a charm and saves lots of valuable time."[23]

Cumnock and his assistant maintained an almost daily flow of letters seeking the myriad of supplies the mills needed, checking prices, and discussing and evaluating quality. They ordered Card Room baskets, firewood, pencils, clutches, thermometers, picker levers, picker sticks, lap sticks, coal, saddles (spinning frame, not horse), rope, bale paper, w.c. paper ("all out"!), hammer handles (machinists' and sledge), harnesses, sperm oil, spindle oil ("winter rendered kidney tallow"), brushes, brooms (and feared rats would eat them in storage when too many arrived at one time), varnish, alcohol, turpentine, burlap, and starch, as well as innumerable machines, parts, and the occasional labor to assemble them when needed. Cumnock also disposed of practically everything left over from the manufacturing process, from iron to cotton waste to machinery. Even the dust from the Picker House was sold "in one horse loads" for fertilizer, as was the stable manure. Time remained for him to write to Pevey Brothers, in New Hampshire, to "call at my office, as I want another cow." They were slow to do so, and three months later he complained: "Haven't heard from you about cow. The one you left here is ugly and I don't want her."[24] (I assume he used the colloquialism indicating "mean" rather than objecting to the cow's looks.)

Cumnock's assistant, J. G. Marshall, wrote many of his letters and aided him in the areas of the job requiring attention but not expertise. For one full workday, Marshall had "been in the mills paying help."[25] It appeared that he freed Cumnock only from such clerical tasks, however.

Cumnock's work, and its toll, were indicated by an episode Marshall described to Treasurer Clarke. Cumnock, he wrote, was "very tired tonight." Mills #1, #2, and #5 had been shut an hour each, and #4 shut since three in the afternoon, because Agent Ludlow, of the Merrimack, put booms across his canal, sending all the snow and ice into "ours by way of the new race way, and if things were to continue thus, they would rather not have water from this sluiceway." Marshall referred to the Boott Penstock, installed by Francis to add excess water from the Merrimack Canal to the Boott's supply from the Eastern Canal. Here, however, these supposedly noncompetitive companies found themselves in conflict as one used the system to serve individual interest at the expense of another. Cumnock, obviously, was in charge of the efforts to keep the mills supplied with power and to deal with the icing, which would clog the trash racks and block the water supply to the turbines. Operating and planning lay equally in his hands.[26]

Cumnock's return for this responsibility may be inferred from a letter to a railroad agent, "Brother Israel." Cumnock's position was such that his son, far from following in his footsteps at local colleges, was at Harvard. In fact, he was captain of the football team. The letter requested a parlor car to take fifty people from Lowell to Springfield, Massachusetts, for the Yale game, and a special engine to bring the car back to Lowell afterward. It implied newfound status, wealth, and involvement in the social life of the city.[27]

Other mills and machine builders turned to Cumnock for advice based on his assessment of equipment. At one point he carried out the inspection of sprinklers from three other mills in order to determine whether or not they suffered from corrosion. When the piston head to the Boott's thirty-two-inch Corliss engine cracked and stopped part of the works, he not only ordered a replacement but also pointed out to its maker that he felt the break could have been prevented by improvement of the proportions of the hub and ribs of the head.[28]

Cumnock's role and knowledge far outstripped that of anyone else associated with the operation. The seat of operational expertise had shifted from Boston to Lowell, at least at the Boott, by the end of the war, rather than by the 1890s, as Lubar suggests. Correspondingly, the Boott treasurers in Cumnock's era appear to have played much less of a role than what was anticipated by Alfred Chandler or Lubar. The changes identified by Lubar for late in the century seem to be in place in the case of this competent agent by 1870. Cumnock's abilities played a major and direct role in the planning, analysis, and operation of the mills on both a day-to-day and long-term basis; the scope of his responsibilities included many important decisions and was much greater than the comparatively passive oversight of self-perpetuating, simple operations that Chandler postulates. Movement toward a variety mill, an order mill, and a finer goods mill all accentuated the extent and importance of his role. The evolution of the operation from a series of comparatively unrelated, independently operating mills to one thoroughly integrated complex further reflected his stamp and magnified his responsibilities and significance.[29]

Limits stood against his action, however. Reviews of the Boott by R. G. Dun, Incorporated, for the last quarter of the century described his reign. In 1876 the field report called it an "A-1 mill," and the next year's report said it "ranks among the best here. Management excellent." Reports continued similarly for a decade, with stock prices ranging from $1,400 to $2,100 (par value being $1,000) and dividends of 4 and 5 percent on investment normal and among the city's best, with a high of 8 percent in 1890. The telling comment came in 1888:

> They had a good year and made money, paid 4 per cent on last dividend. It is the general opinion that they do not keep their plant in as good condition as they ought, and in this respect at least the management is thought injudicious, necessitating a large expense at some time. They are in high credit and standing.[30]

Dividends and stock value flowed from effective operation of the plant, which included the choice and operation of the machinery, and execution, areas basically in Cumnock's hands. Furthermore, the lack of attention to the plant's maintenance did not damage the mill's standing, since this style of operation was traditional in Lowell and served the stockholders' immediate interests. On the other hand, after Cumnock left the Boott in 1898, he and his in-laws bought the Appleton Corporation, which he ran as treasurer. He rebuilt it from near-bankruptcy, and it was later said that after twenty-one years "not a building stood there that was there when he took control."[31] If Cumnock operated in that manner when in the seat of power, it suggests that it was previously the stance of the directors that produced the "injudicious" policy described by the R. G. Dun reporter. The decisions most important to the future of the operation remained the purview of directors with a less than far-sighted point of view, at best, and a willingness to run the mill into the ground for short-term advantage at worst.

Technological Choices

Equipping a New Mill

Detailed records of the equipping of Mill #6 in 1872–73 offer a picture of the thought process behind the selection of cotton technology at the time. For power, the Swain Turbine Company had the Lowell Machine Shop (LMS) build an 80" wheel weighing 9,409.5 pounds. With regulating gates, bracing, gearing, and shafts to carry power from the turbine pit in Mill #1 to #6, it cost $18,853, delivered and installed.

In January 1873, the mill received the two crown gears ($959.64), which would take power from the wheel, and in February the shop sent the "Double-Acting Regulator," which would govern the wheel's speed (by controlling water flow) so that it and the production machines would run at a steady pace no matter what equipment was turned on or off at any moment. Finally, shafts and gears "to connect #1 & 2 with #6 Mill": power could now flow from the wheel in #1 to the new #6![32] Thus were readied the engines, for water and steam both drove engines providing the motive power for the new factory. In effect, the mill was one great machine with an outside power source.

The Carding Room contained 152 cards, united by 8 railway heads, fed 8 drawing heads, 4 "coarse compressed roving speeders" (96 spindles), then 6 intermediate speeders (288 spindles), and finally 14 fine speeders (1,008 spindles).[33] This group of machines occupied the top floor of the building, 6–5, despite the weight of the carding machines. Their calm rotary motion contrasted so sharply with the violent action of the looms that they were better located high in the structure. Further, placing the structure-threatening looms low in the building necessitated putting the cards at the top to avoid a convoluted flow of materials. In this set-up, cotton moved from the Picker House to the top floor of #6 in the form of laps (batting) via the elevators, then down through the mill until it left as cloth, much like the flow of grain into flour in the prototypical Oliver Evans grist mill.

Seventy-six breaker cards took the picked cotton and separated the clumps of fibers one from another while beginning the parallelization process, which would continue until the cotton was spun into yarn. In each card, a large main cylinder, held in a strong and rigid frame, spun rapidly as the cotton was carried around on it and between it and the fixed top-flats; both cylinder and flats were covered with card clothing, and this wire-brush-like material held and straightened the fibers.

The cotton laps thus produced then went to 76 finisher cards to increase the untangling and parallelizing effects. The care and precision employed here affected all operations to follow, making this equipment very important to the quality of the mill's output. The row upon row of evenly spaced cards stood, covered, constantly devouring and emitting streams of white fiber, 152 islands of iron and wood amidst a sea of cotton.

Mill #6 began operation about ten years after the widespread adoption of George Wellman's top-stripping apparatus (invented in 1854 by an employee of the neighboring Merrimack Manufacturing Company), which significantly affected card-room work and efficiency. Wellman's device automatically cleaned the top flats of the card of the dirt and waste that accumulated on them, which previously had to be removed by hand. Costing $60 per card, the device was said to have saved $300-worth of labor annually.[34] By installing these strippers, Boott managers made their card room technically current in the American field.

The railway and drawing heads took the sliver, or ropelike product of the finisher cards, combined a number of them, and then drafted, or drew them out, enough so that the new product in each case was as small or light as any one of the slivers that had entered the machine. These doublings represented the first of the thousands to be employed in order to compensate for the imperfections of the carding. The drawing heads lifted the slivers from the tall cans into which they had been coiled. At the next stage, the cotton began to assume a more yarnlike shape and was wound onto bobbins. Here, the use of coarse, intermediate, and fine speeders represented the best solution available from the LMS at the time. For United States practice, using the three machines stood as the preferred style, but it did not equal the abilities of the English fly-frames which were then coming into the market and which would soon become dominant. The deficiencies would become increasingly apparent as the Boott shifted to more varied and finer production.

The three types of speeders stood in long rows, creating aisles through which workers and cotton moved. The coarse speeders drew sliver from cans, doubled and drafted them, producing roving, the greatly reduced product of this speeder. The bobbins of roving then went to the second and third speeders, which repeated the process.

Each of these speeders wound the roving onto the bobbin by means of a flyer, or throstle, a device much like the flyer of a spinning wheel, which revolved rapidly around the bobbin and made a noise compared to a thrush. The room full of speeders would, of course, sound like a great flock of thrushes accompanied by a cacophony of noises associated with the belts, shafts, pulleys, gears, hand-trucks, and employees moving through its aisles.

These fifth-floor operations transformed the cotton from laps, or sheets, to sliver, to narrow rovings, perhaps between half the size of a pencil and pencil-sized. Trucks of these packages of roving on bobbins constantly moved through the room to the elevators, where they descended to the spinning department below.

The spinning equipment was of two types: ring-frames and spinning mules. Thirty-six ring-frames, each with 224 spindles spaced just 1 5/8 inches apart, created aisles much more dense with spinning bobbins than those on the floor above and also turned the bobbins much faster, on the order of 6,000 revolutions per minute (rpm). In spinning, #6 arrived at the end of an era: in 1870, for the first time, the LMS made no throstle spinners, thus ending a transitional period, begun in 1855, during which they offered both rings and throstles.[35] Therefore, the decision to install ring-frames in #6 was an easy one.

However, in just those years around 1872, revolutionary developments were taking place in the technology of spindles, the all-important parts that spun the bobbins, and the managers of the Boott faced complex choices between the tried and true and a variety of new inventions. At a meeting of the New England Cotton Manufacturers Association in 1873, Cumnock described the process and considerations that led to his determination of which spindle to acquire. His comments indicated the extent to which mill operations were carried out according to the best knowledge of the agent in charge; they underlined the degree to which constant efforts in the mills evaluated and extended available knowledge; they described the source of technological innovation; finally, they revealed the extent to which the corporations shared their discoveries, debating them as they did here in a public forum, confident that their several companies did not compete significantly with one another and therefore did not benefit from the harboring of secrets. The men at the bottom of the financial hierarchy but the top of the operations ladder, the true cotton experts, recognized that they in effect all worked for the same small group of investors who gained from their sharing knowledge.

The new spindles becoming available were lighter and could run faster with less power. Previously, however, Cumnock's experience had been that "the girls could tend 1,120 of the heavy spindle, to 960 of the light." When he determined to test the new spindles despite his prejudice, he encountered another problem: "Before I commenced the experiments on the new styles of spindles I had my doubts of getting a fair and correct result, on account of the bias of overseers and the influence of patentees upon them." He had "friend Richardson," LMS superintendent, build a frame with a light spindle, which he then ran in the plant without anyone being informed as to its difference. He also had "Mr. Draper," an inventor and machine-builder, adapt a frame to his spindle, and he ran that similarly in order to test performance and power consumption, avoiding the results of "a test set up for the occasion, as I fear a great many of them have been that were sent us."[36]

Cumnock's statement represented a tour de force of technological sophistication. It revealed a technical knowledge and precision encompassing experimentation, measurement (of both speed and power), and evaluation (of both performance and effect on labor). Cumnock went on to describe the continuing assessment of the spindles in #6, justifying its extra cost through the efficiency of its performance. His discussion indicated the amount of effort, and expertise, applied to the selection of what could appear to be a standard assortment of equipment.[37]

Yet despite all these words of explanation, a reader might still respond, "But it's only a spindle. Why all the fuss?" Spindles inserted the twist into the yarn as it was being made. Therefore their speed presented a limit on production; increasing their speed had the potential to speed the entire mill. Decreasing their weight, and thereby power consumption, saved money.

It is worth noting that the Wellman card stripper and the Sawyer spindle, as well as competing spindles, were the products of experimentation of those who worked in the mills rather than of inventors in machine-building shops. Little innovation had been taking place in recent years: "The cotton-textile machinery industry of the 1870s had been softened by three or more decades of heavy demand. In those decades the advance in machine technology had been very slow. On every hand there were complaints from mill men that machinery companies had not changed their models in thirty years."[38] (Thus the failure to replace speeders with fly-frames.) When new ideas did appear, they arose from the shop floor, an ominous sign for the future of the quiescent machine builders.

In addition to the 8,064 ring spindles purchased, twelve mules of 648 spindles apiece (7,776 in all) were also installed on the third and fourth floors, which, with their one centered row of columns, were ideally suited to accommodate these giant machines. Each mule was on the order of seventy-five feet long with a moving carriage full of spindles rolling back and forth through an approximately five-foot draw. Viewed from the end of the floor, they ran in facing pairs, the spindles of the mules separated by only a small walkway when the carriages of two machines were in their "out" position, by about ten feet if both carriages ran in (toward the creeled rovings about to be spun). The back or creel of one mule lay to the windows on the canal, that of the next mule facing it toward the center posts; behind that, on the other side of the posts, stood the back of another mule facing a fourth with its back to the windows on the yard side. The view of the machines in operation showed all these carriages, rolling to and fro, each spinning 648 spindles, first running out from the creel as the roving was stretched and twisted, then running back as it "wound on" the yarn just made—an expanse of 648 parallel pieces of yarn five feet long diminishing to a scant few inches after the rest was wound onto the bobbin, yarn accordions constantly expanding and contracting.

The LMS also provided 285 looms, each capable of weaving 40-inch-wide cloth and running at 135 picks per minute (ppm). These were arrayed in rows on the second and third floors. Placed in facing lines, they left little space between them for the weaver to work in, 1'9", in fact, while behind them, where access was required for bringing in bulky warp beams, wider aisles remained.[39] Simple plain-cloth looms, their thundering operation resounded through the building, shaking the floors and walls. Their limited capabilities did not suggest a long life for them when the mill began to vary its production.

This carding, drawing, roving, spinning, and weaving equipment completed the basic productive capacity of Mill #6 and cost $151,982. In general, choices primarily required consideration of productive capacities and coordination of numbers of cards to spinners to looms, since they were basic units of production for the time for a mill relying on the LMS. When there were other considerations, however, as in the case of spindles, for which choices were not automatic at the time, great care was taken.

Of course, running this operation demanded innumerable related tools and supplies which did not appear in inventories of major purchases. These small items had to be ready at all times if the juggernaut of a factory were not to be halted for want of a nail, as in the horseshoe-horse-battle routine. While the LMS provided the big items, numerous small shops had arisen in and around Lowell to fill these sundry needs.

Mill #6 also consolidated the mill complex's service facilities—Machine, Paint, Carpenter, Blacksmith, and Belt shops, on the first two floors for all the mills. The machine shop equipment of various vintages was assembled from other locations at the Boott. With it, machinists could make numerous and substantial repairs on the textile equipment. New gears could be cut to replace those that broke or wore out. Lathes could hold pieces of metal up to twenty-five feet long and turn out shafts, cranks, or other round parts. The Daniel's planer smoothed large timbers, its whirling knife-tipped arms a pronounced danger to operator or passerby alike. Wood-turning lathes and saws permitted repair of wooden machine- and building-parts. The presence of this large group of machine tools made the mill comparatively independent of outside repair facilities. It indicated the amount of wear and breakage that could be anticipated, and, through the size of the investment it required, the importance of ready repair to the mill's success. The variety of equipment could work on the smallest and largest parts of the system, from machines to the shafting system, and even the buildings. The Blacksmith Shop could bend parts into needed shapes, could harden the steel which had to bite against softer metal, and could weld together those parts which were not bolted or screwed. The role of the Paint and Carpenter shops is parallel and more obvious, but the connection between the Belt Shop and the rest of the facility may be less clear.

There was another part of this system, this factory-as-machine: the complex array of shafts, belts, and pulleys which brought the power from the engines, water and steam, to the building, the floors, and the machines. No machine ran off an individual motor. Rather, whole buildings of machines ran off one or two motors. How this system was structured and operated had much to do with the shape, layout, and performance of the mill and its equipment.

The installation of the Swain turbine represented the beginning of the power system that ran Mill #6. The shaft of the turbine rotated in a vertical plane, and two bevel gears turned that power 90 degrees so that it ran horizontally through another shaft. This shaft ran parallel to #6 until it reached a point from which a leather belt could be run from a large pulley on it to a similar pulley in #6-1. The belt connecting these two pulleys would have been on the order of twenty inches wide or more, made of several thicknesses of leather, and in its entirety consuming numerous cowhides. Once the power reached #6, it had to be carried up to each of the five floors before it could be directed to the many machines. This transmission paralleled that from the wheel to the building, but in a vertical plane, with belts only slightly smaller than the first revolving around a pulley on the first-floor shaft and another on a shaft on at least every other floor above. Once the shafts on given floors were turning, power could be taken from them down to machines below, or through the ceiling/floor to machines above, by smaller belts, to a pulley on each machine. In one sense, this was a simple system, all mechanical, no electric motors, computers, or gear boxes. At the same time, it was a highly sophisticated, potentially efficient system with a thorough theoretical grounding by the time of its installation in #6.

Cumnock, in an 1879 comment, added a refinement on the question of shafting and loom speed; in order to reduce the extreme vibration problem which these machines presented, he called for both balancing the shafting to minimize its vibration and running alternate rows of looms at plus or minus 3 ppm to help keep them out of synchronization.[40] The looms were the machines most threatening to the structure. Each loom threw a shuttle and bobbin weighing pounds back and forth across its width 135 times per minute. This action created serious vibrations at all times. If all the looms on a floor coincidentally began to perform this action simultaneously in a particular direction, in other words if all the shuttles started flying first north and then south at the same time, the swaying action created would be tremendous, and mills had been brought down by such coincidences. Cumnock's precaution reduced the danger.

In picturing any part of the mill, one must keep in mind the omnipresence of lines of shafting running down the ceilings of the rooms, with belts running from the many pulleys (equal in number to the number of machines) to every machine. On the machine, two pulleys sat side by side, one the idler, loose on its shaft, the other the fixed pulley, fastened to the shaft. When the shafting system was turning, a belt running to an idler turned but did not operate the machine; but when an operative pushed a lever moving the belt over onto the fixed pulley, the machine started up. In most cases these levers, known as shipper handles, were on the machine, but in the Machine Shop the two pulleys in question were located on the overhead shafts and drive belts ran to fixed pulleys on the lathes and other machines. The machinists turned the power to their machines on and off by means of long wooden handles extending up to the shaft above their heads. It is indicative of the pervasiveness of this power system that these handles created the appearance of such a forest that the machine shop foreman was invariably known as "the Bull of the Woods." Even the general language of the time, outside the factory, shared in such common usage, and a person without full occupation was said "to be on the loose pulley."

The shafts, pulleys, and belts, along with the groups of machines described earlier, made Mill #6 one great machine, interconnected by purpose as well as literally, by metal and leather. At the same time, it was an integral part of the complex. Other sectors of the mill served #6, providing picked cotton, power, heat, and other services. Mill #6 served the rest of the complex in turn, through its several shops for machining, painting, carpentry, belting, and blacksmithing.

The planning of the production line and the selection of its components reveal careful and informed thought. Given the mill's goals, determined by the investors and the treasurer, the machinery selected indicated full awareness of its capabilities. The openness of the exchange of information which led to the choices described reflects operating conditions and reveals the limited nature of competition among the manufacturers. Only one note introduced a discordant sound to an otherwise harmonious situation: the fact that although the LMS assumed, correctly, that it would equip these mills, the technological innovations were coming from people outside its employ, on the floors of the mills. Complacency, on the part of either the mill or the Machine Shop, carried the threat of relative decline if others worked more assiduously to advance. Whether or not such a threat could be met by the combined strength of these textile giants remained to be seen.

Changing Machinery

Equipment installed between 1871 and 1904 provided a good indication of the intentions and performance of the mill's managers. While it is axiomatic in certain schools of architecture that form follows function, in a textile mill function follows equipment, and they must change together. The Boott could choose to continue with existing equipment, replace it with newer versions of the same general type, or purchase machinery with new capabilities, signaling new intentions. All these things happened, and in various ways they not only suggested management's plans but also redefined general operations and workers' practices within it. In fact, the choices demonstrated a furthering of longtime practice, the search for automaticity, and a new goal, the capacity for finer and more varied production.

View of the mills from *Gleason's Pictorial Drawing Room Companion*, 1852, done from inside the brick wall enclosing the complex, shows Mills #3 (*left*), #2 (*right*), and #5 (*to rear*). Smaller Picker Houses, *in front*, separated this process, and its fire danger, from the mills. Courtesy the Museum of American Textile History.

Employees pose in and around millyard, now (ca. 1878) enclosed by buildings. Courtesy the Lowell Museum.

Overview from insurance survey in 1882. The Eastern Canal lies in front of the mills, the Merrimack River beyond. Courtesy the Museum of American Textile History.

The millyard in 1893, with a fenced lawn in place of the plantings of an earlier day. Flat roofs on all buildings and the absence of the ornate houses for hydrants and firehoses lends a strictly functional appearance to the scene. Courtesy the Museum of American Textile History.

The plant during the 1930s Depression, with the Counting House (*left*) and Mill #6 behind the departing figure. Courtesy the Museum of American Textile History.

In the 1950s, pavement replaced lawn, as did trucks the horses and railcars of an earlier era. Courtesy the Flather Collection, Lowell Museum.

VARIETY

The Boott did not replace its cards with the greatly superior revolving-flat cards available early in this period. The result was a loss of efficiency relative to competing mills, and a loss of quality as well, particularly regarding finer production. Outmoded railway heads and drawing frames also remained in place. The Lowell slubbers (or coarse speeders) purchased as late as 1889, while gearing up for finer and more varied goods, represented an error in judgment. The new English fly-frames with which the Boott replaced most of its intermediate and fine speeders during this period served particularly well in producing the finer yarns toward which Boott moved. Unlike the speeders, which could only be changed from one count to another through "intricate adjustment of bobbin and spindle,"[41] the new machines offered truly flexible operation. George Sweet Gibb described the situation:

> Not only did the speeder require intricate adjustments, but it failed to impart enough twist to the roving. On coarse work this latter deficiency was not important, because the roving was strong enough to be processed, even with little twist. On finer numbers twist had to be imparted to give the roving sufficient strength to hold together under subsequent processing.

In spinning, Cumnock's astuteness in choosing the Sawyer-spindle ring-frames in 1872 assured that little change was necessary in that department. More spindles of ring-spinning (with similar Rabbeth spindles) were added. The capacity for mule-spinning, however, increased still more rapidly, again reflecting the move toward increased quality and flexibility, the mule's fortes.

The mules made a softer, loftier yarn than the ring-frames; they produced the weft for the looms. The strong, harder yarn from the frames made up the warps. The mules were a high-skill machine, and for that reason they were avoided in Waltham and early Lowell. Their efficiency, flexibility, and quality of product had partially overcome those feelings and they were introduced at the Boott in 1849. Their presence indicated a realization by the Boston Associates that their plans to eliminate skilled labor had limitations. Over the years the ratio of mule- to ring-spindles gradually increased, and during the 1880s still more of the former were purchased, making the proportions nearly equal.[42]

Capable of producing finer yarns as well as being more easily changed from one weight and twist to another than spinning frames, these machines were eminently suited to a company interested in producing varied, finer products. In fact, altering both the amount of twist inserted by the mule into the yarn and the amount of draft required only the relocation of two indicator pins, a change so easily made that the pins were sometimes protected by locked covers in order to prevent the mulespinners from changing the settings to suit their own fancies (which might have more to do with achieving high production and thus high piecework wages than matching predetermined standards).

The Boott's ambivalence toward the new equipment and the mule-spinners appeared in their annual listing in the statistics of production where "Males Employed" included the notation, "Mule tenders" (no other occupation being noted). Their presence in the mill changed the mix of employee skill levels significantly, just as their machines' altered production capabilities. Shifts from the mill's original technology represented, to the owners, one of the prices entailed by their shifting production strategies, particularly in the eighties and nineties.

Looms, however, represented the technology most often and thoroughly revised during this period, and the reasons stemmed from both changes in available technology and new production goals. First came LMS looms that incorporated some of the technical advances of the 1870s and 1880s such as the Bartlett let-off (on the warp beam), the Estes, Amoskeag, or Boott take-ups (on the cloth roll), and the Stearns parallel motion (on the picker stick).[43] In each of these cases, the new machines displayed the results of continuing effort to make power weaving more consistent, smoother, and less labor-intensive. Such innovations were constant in loom engineering and indicated the imperfection of all methods in the difficult task of combining the varied motions that go into weaving.

Adding Crompton and Knowles fancy looms brought more significant change. They manipulated up to twenty harnesses to produce complex styles of cloth. These looms represented a change not of degree but of kind in the weaving operation. Technology, labor, and product were all significantly affected.

Instead of using cams turning over treadles beneath the warp to create the shed through which the shuttle passed, these looms employed dobby heads on the side of the loom arch to control the harnesses. These mechanisms consisted of a series of levers to lift or depress each harness for each pick of weft. They were programmed by an endless chain (a loop) to produce a predetermined pattern.[44] Not only could these mechanisms control many more harnesses than could cams (on the order of 20:6), but they could also be easily changed to produce a variety of fabrics. New chains, with a different arrangement of risers and sinkers, needed simply to be prepared, away from the loom, and exchanged at the time of a style change.

These looms also incorporated drop-boxes, which permitted the weaving of various weft colors. Several shuttles, carrying bobbins of different colors, were held in boxes on either side of the loom, with one box always empty to receive the shuttle last thrown. Thus a 4 × 4 loom included four shuttle boxes on each end of the lay and could weave with up to seven weft colors.[45] These looms, by their combination of complex and easily changeable warp and weft control, permitted the introduction of new complex styles and facilitated shorter production runs. These changes increased both the variety of materials and the variety of production problems which the mill had to face.

These new looms (and fabrics) also presented new situations to everyone associated with their operation. Managers had to order production of a greater variety of raw and finished materials within the mills; spinners had to provide quality yarns to withstand their increased manipulation by the looms; weavers had to develop greater skills to maintain higher standards of production and to avoid faults in the more valuable product; loom-fixers had to learn new and more precise repair techniques; even the marketing of the mill's output had to shift. The fabrics now produced suddenly thrust all elements into a new and more complex arena of operations.

These looms, more than any other single change in the machinery of the Boott, suggested the extent of the reprogramming going on at this time. They were not installed because they could run faster and produce more cloth, but rather because they could make a more complex cloth, as well as a wider variety of cloth, and thus more valuable products. In other words, they offered the hope of increasing profits not by expanding and speeding up production of the traditional coarse and unchanging materials, but by generating smaller amounts of complex, changing, and more highly priced cloth.

AUTOMATICITY

New Draper looms also appeared late in the period. Although they were cam looms like those they replaced, that similarity is misleading. For one thing, these looms wove corduroy, the familiar fabric with weft threads manipulated to form the loops of a pile. The loops were later cut to create the brushy, soft surface associated with this cloth. So, to begin with, the new Drapers introduced another new and more complex material.

Yet that change paled in comparison to the real impact of these machines. Because of one major change in how they operated, these machines were known as "automatic" looms from the moment of their introduction in 1895. In weaving on all earlier looms, replacing the nearly spent bobbin in the shuttle required steady attention from the weaver and acted as the primary limiting factor on the number of looms that one person could tend. Since weaving entailed roughly half the labor cost of making cloth, continual efforts had been made to overcome this limitation through devices capable of replenishing the weft without the weaver's assistance. Most of these devices involved shuttle-changing mechanisms; none was successful.

In 1889 the Draper Loom Company of Hopedale, Massachusetts, had hired James H. Northrup to lead an effort to overcome this problem. He and his coworkers greatly affected the development of textile history when they invented a series of mechanisms which would automatically change the bobbin, rather than the shuttle, while the loom ran.

The key features of these looms included a circular hopper, or "battery" (like a Gatling gun's), mounted on the right-hand end of the loom and holding a supply of full bobbins; a filling detection system; and warp and weft stop-motions. Each time the shuttle reached the left side of the loom, a feeler reached in through a hole in the side of the shuttle and felt the amount of filling left at the base of the bobbin. If the supply had nearly run out, at the next pick the empty bobbin was knocked out of the shuttle through a hole in the bottom and a new bobbin was inserted from the battery, all without stopping the loom or any intervention by the weaver. Because the battery took over such a time-consuming part of the weaver's work, a worker could tend many more looms. Since the person's oversight of the larger number of machines could not be so close as it had been before, other mechanisms were required in order to take advantage of the loom's capability. For example, a new weft stop-motion detected a broken weft yarn, and a new warp stop-motion noted a broken warp thread; either immediately stopped the loom.[46] The need for the new stop-motions suggests that earlier versions, cited as justifications for stretch-outs by some, had been ineffective.

These looms represented a quantum jump in weaving technology, and the Boott was very quick to seize upon this opportunity to significantly improve a part of its equipment. In fact, it was one of the first mills in New England to use them. Boott management adopted the "ideal" weaving technology more quickly than most other operations of its time by buying Draper automatics in the first years of the century.

Chapter Four

The Interaction of People and Machines

JUST AS THE SYSTEM of shafting brought power to the machines, so did the workers bring the machines to life through their attendance at them. They changed the scene from one of hardware to one of activity, and the intensity of this activity had grown fierce.

Each room had its own varieties of work, with a predominant mode as well as a plethora of associated tasks. In the Card Room, workers carried or trucked the large laps of cotton, about four feet long and some two feet in diameter, wrapped around protruding lap sticks, to the feed of the breaker card, then from that card's delivery to the feed of a finisher. The sliver from the finishers was carried by the mechanism of the railway head into a tall (3'–4') tin can which had to be removed when full and taken to the drawing frames. This task was repeated to get the fiber to the coarse speeder, after which the roving was wound onto bobbins as it progressed. At the intermediate and fine speeders, a worker would doff, or remove, bobbins as they became full and replace them with empty ones. The full bobbins would be placed in trucks that other workers would roll to the elevators and take to the spinning department.

The work described moved the cotton through the first department, but it only began to account for the activity there. Different workers cleaned the carding machines, oiled them, repaired and re-clothed them with the wiry card clothing. They ground, or sharpened, the card clothing, mounting emery wheels on the card frame to do so. Sweepers and scrubbers cleaned the floors. The contributions of all these people were immediately necessary.

Work in this room was the dirtiest in the mill. "Fly" filled the air. This loose cotton and dust, now recognized as the cause of the debilitating byssinosis, or "brown lung" disease, permanently diminished the breathing capacity of many cotton mill workers. Because of the dirtiness and the heavy lifting involving laps, along with skill in certain jobs, most of the people actually involved in carding were male, though in the same room women tended the speeders and fly-frames.

The hierarchy in ring-spinning paralleled that in carding. Workers doffed and replaced full bobbins, reattached broken ends, and picked off threads errantly winding around the drafting rolls. Others supervised, maintained, and repaired these frames, trucked the product, and cleaned. Work was less dirty than in carding, less heavy, but fatiguing from its constant nature, the stress of piecework employment, and the hot and humid conditions pervading the buildings. Women tended the frames, their jobs defined by the number of "sides" minded: they moved up and down one alley between several frames rather than moving around each frame.

The situation in the Mule Room was different. There the chief attendant for each pair of machines, the mulespinner, enjoyed considerable status. He controlled his helpers, piecers, who reconnected broken ends (of yarn) as the machine ran and doffed it, stopped, when the bobbins filled up. Despite the fact that these machines were said to be automatic, or "self-acting," mules, his skill remained considerable. He was hard to replace, the prime measure of worth in the mill's account books, and therefore well-paid. Work here was male, somewhat controlled by the power of these men who found their own assistants and regulated admittance to their trade. Hot and humid conditions helped the yarns to be formed most efficiently, and so the mulespinners generally wore their pants rolled up, wore light shirts or none, and worked barefoot. Boys not only brought fresh roving to the machine's creel, but also moved underneath the yarn to pick up broken ends and clean up the fly accumulating there. The machine was stopped for this operation; a failure of communication (the boy threw up his hand as he emerged) or inattention would result in the long carriage running in before the boy was out, catching and crushing him against the back beam. Danger from byssinosis and other occupational hazards (e.g., testicular cancer from the spindle oil) also persisted here, as they did, in fact, throughout the building.

Following the path of production, the bobbins from the mules were trucked to the weave rooms, where they would be placed in shuttles and provide the weft for the various cloths. Work there followed the same pattern as on the floors above: all the supervisory and more skilled jobs, such as loom-fixer, were the province of males, with women in machine-tending and cleaning roles. Since the work of the Weave Room was greatly affected by the changes during this period, detailed discussion of it follows in conjunction with the technological causes.

EFFECTS OF NEW MACHINERY

The impacts of technological change were very unevenly distributed during the period 1871–1904. Card Room employees operated the same machines at the end of the period they had at the start. Changes in roving machinery had little effect on the work. In spinning, the classic effects of the speed-up and stretch-out appeared more obviously. The Sawyer spindles' increased speed raised production. Changes in numbers of spindles and workers offer gross comparisons between loads over time in this area. In 1866, 1,020 women worked at the Boott while 71,300 spindles spun; in 1871, one finds the same number of female workers and 92,000 spindles, and by 1883, an operation with 140,000 spindles employed 1,400 women, less than a third more women for twice the spindles of seventeen years before, despite the fact that faster production required increased attendance by labor, particularly in doffing. Since the increase in male employees, from 400 to 475, in the time was still smaller, it only heightens the apparent effect of the stretch-out. For the mulespinners, there were larger machines, more spindles to tend, thus more workers to oversee and greater distances to walk while running the longer machines. The work itself had altered in terms of the yarn's variety and fineness and in the difficulty of making it.

William Burke, a former Boott agent, offered his view on the comparison between work and workers in weaving. In 1876, thirty-two weavers tended 194 looms, or just over six apiece, in contrast to two apiece in 1838. Overall, each operative in the mill produced 3.33 pounds per hour. He also noted that whereas in 1838 11 percent signed the payroll with marks (indicating illiteracy), in 1876 25 percent did. Health had improved following the change from a 76-hour work week to the 60-hour week of this time; the incidence of "bowel complaints, dysentery, typhoid, and fevers" particularly during August and September had declined, perhaps because the operatives were able to eat breakfast before going to work, in his opinion.[1] Without significant technological change, by 1876 the mills had reduced hours but raised loads, or stretched out labor, and increased production, and anticipated healthier, if less educated, workers.

The increase in number of looms, from 1,880 in 1871 to 3,889 in 1883, before the advent of the Draper automatics, represented another major stretch-out, since the number a weaver could tend had not been greatly altered by technology. Therefore, this large increase in looms magnified the effect of the stretch-out elsewhere in the mill, increasing the effects calculated above. For example, if each weaver tended 8 looms, a very heavy load, 251 workers would have been required for the additional 2,009 looms, and fewer of the new workers would have been available for spinning or other processes producing at the rate required to supply all these looms. Production nearly doubled in this period, from 380,000 to 700,000 yards (of lighter cloth) per week. Loads increased dramatically, despite the fact that this was not a period in which major labor-saving devices were being introduced.[2]

The weavers' work bore the brunt and occasional benefit of technological change at the end of this period. In addition to the innovations already mentioned regarding the Draper loom, the new loom also required a new self-threading shuttle. In these a clamp grasped the base of the bobbin; when the bobbin emptied, it was forced out and a full bobbin was pushed down into the clamp; the yarn on the full bobbin slid down into a self-threading slot near the nose of the shuttle. The significance of this aspect of the new loom lay in the realm of worker health. The fastest way to draw the thread through the ceramic eye in the side of the old-style shuttle was with a quick, sucking "kiss." If the person who performed this operation had a communicable disease such as tuberculosis, the second weaver was in line to contract it. In any case, lint, dyestuffs, and other foreign matter were inhaled, hence the name "kiss-of-death" shuttle. It is no wonder, then, that 70 percent of textile operatives died of respiratory disease at a time when only 4 percent of Massachusetts farmers died from this cause.[3] The new loom incidentally eliminated this hazard.

A weaver still had to accomplish most of the tasks described in Chapter 1, only now some were signalled and weft replenishment was automatic. The stress involved in trying to keep a full assignment of looms operating, whether eight or twenty-five, can be pictured as the weaver moved from loom to loom either because they have shut off or because they are about to, with the knowledge that every idle moment of loom time represented a diminution of pay. The automatic looms, while accomplishing weft replenishment on their own, led to roughly threefold increases in loom assignments, requiring a weaver to make repairs to weft and warp breaks on some twenty-five looms, with a 100-foot corridor of looms demanding attention. The pressures that had resulted from increases in assignment earlier in the period were multiplied as the company sought to enjoy the benefits of the new technology.

The new patterns of production, moving to finer and more various goods, and fulfilling orders rather than producing for inventory, caused changes in the management of technology, with profound effects on all parties involved. For workers, the effects included shifting between changing products, keeping track of the new products, coping with the vagaries of raw material supply caused by such variety, dealing with the higher standards required for finer products, as well as the increased likelihood of problems associated with the lighter, weaker materials, whether roving, yarn, or cloth. Each of these effects increased skill requirements and extended pressure to produce. Carding had to be performed to more exacting standards on more varied cotton to produce changing weights of sliver. Lighter roving broke more often and required careful piecing to avoid lumps, or slubs, which were less acceptable than in coarse goods. Correlation of carded material provided and roving to be produced also had to be exactly maintained. Spinners faced parallel difficulties, as did all employees. For weavers, particularly, the changes magnified the skill and attention required. It was harder to see and to tie a knot in a finer warp thread, which was also more likely to break and was harder to draw in through several harnesses. Lack of care in stopping and starting the loom would be more quickly revealed, and that or any other error would cause a greater loss in value to the product and be more heavily penalized. The conjunction of the several factors involving fineness, variety, and shorter runs, particularly in the eighties and nineties, multiplied the demands on the workers. Employees tending more and faster-running machines had to demonstrate the flexibility to deal with rapidly changing materials and products.

Effects of New Procedures

Procedures in the mill revealed another aspect of its technology: the ways workers were organized to participate in it. English observer T. M. Young described the operation of the "tackler," or loom-fixer, and a "room girl," in the Merrimack Mill, literally next door to the Boott and in all ways a very similar mill. He had been surprised at the number of looms assigned to each operative until he observed the replacing of a warp beam. Before the warp ran out, another was brought in on a hand truck carrying four vertically:

> As soon as the last cut was finished the tackler and a woman specially employed for such work took the loom over from the weaver. The woman helped the tackler to "gait" each beam, and in fifteen minutes the whole job was done and the loom running again. The weaver had nothing to do with the new warp until the loom had been started and was working smoothly.

Similarly, the weaver did not doff his or her own cloth; a "room girl" did it and reported any serious faults to the overseer, and "the weaver would be 'called up' before the whole room. This 'calling-up' is felt by the weavers to be something of a disgrace, and the superintendent assured me that its disciplinary effect was even greater than that of the fines."[4] The descriptions, and Young's surprise at the procedures, underline the extreme subdivision of work and the careful manipulation of labor in the Lowell system. The loom-fixer, a more skilled worker, played his limited role assisted wherever possible by cheaper help; to this was added a system to embarrass a weaver with "serious faults" in a piece. Each part of the system minimized the role, and dignity, of any individual. The mill's need for knowledgeable fixers combined with their organization to make them among the most able to resist the pressures of the system. The lesser skill of the weavers subjected them to piecework and ridicule in a continuing effort to instill the new morality and to squeeze a last bit of production out of them.

Another procedure, oiling, while seemingly minor in itself, reveals the astuteness and aims of the technology's managers. The significance of lubrication for the maintenance of the machinery had long been recognized. Cumnock in 1868 described the system devised at a mill in Nashua, New Hampshire, and adopted by the Boott under Burke, his predecessor. Instead of placing a 10-gallon can in each room for the workers to draw from, they used one 1,500- and one 600-gallon tank. A man then filled individual oil cans, tested them (for both performance and leaks), and delivered them to each room on a daily or weekly basis. He then "pass[ed] to each hand *his* or *her* oiler" and received the used one in return.[5] This system kept leaking oil cans out of the mill, monitored the use of oil, and insured that the workers were prepared for necessary oiling every day. It typified the degree of specialization and routinization applied.

Cumnock indicated the importance of this seemingly minor aspect of the mill's operation in 1879. After tests at the Massachusetts Institute of Technology showed that his oil had excess "gum" in it, he switched to another with equal resistance to evaporation and as high a flash point *and* gained 8 percent in power.[6] This attention to detail indicated the careful study given to all aspects of the mills' operation and the degree to which the approach was made scientific, and it offered partial explanation of his high reputation: "When we personally first knew Mr. Cumnock he was agent of the Boott Mills in Lowell, and the Boott for a great many years, under Mr. Cumnock, seemed a great success. Mr. Cumnock went out, and almost immediately the Boott became close to being a disastrous failure."[7] Given the innumerable opportunities for variation in the performance of people and machinery in a cotton mill, the importance of Cumnock's knowledge, skill, study, and dedication stand out.

Fire danger in the mills brought similar attention. The potential for sparks was plentiful, particularly in picking, with volumes of cotton dispersed in the enclosed space and rapidly moving metal parts available to strike a spark on any foreign matter. Picker houses separated this process from the rest of the mill structurally early in the industry's development. Throughout the mill, however, cotton offered a ready source of fuel should a fire begin. Danger from oil-soaked floors, overheated bearings, accumulations of waste, and oil- or gas-lamps offered a constant threat of a fire capable of consuming a mill with great rapidity.

Both the Locks and Canals and the Associated Factory Mutual Fire Insurance companies conducted quarterly inspections of the buildings and yard at the Boott to check the condition of sprinklers, hydrants, extinguishing apparatus, and the rooms themselves, as well as certain employee conduct. Regulations spelled out the duties of the people involved throughout the mill and yard (watchmen, overseers, agent, and superintendent, and Locks and Canals employees to deal with the water-supply system). The regulations were noteworthy not only for their account of procedures but also for their specificity regarding conduct. Watchmen, for example, not only had to be present but also had to "drive watch-clock pins" hourly from "ringing out" at night until work resumed, and at any other time the mill was idle, whether lunchtime or holidays. This practice involved carrying a clock to a series of stations around the mill and using the pins, or keys, at those locations to mark a recording paper in the clock, thus proving that the circuit was being made, and at the times set. If they found a fire, they were to fight it, but as soon as it appeared impossible to overcome the watchman had to sound the alarm and immediately turn on the sprinklers to flood the floor involved. All iron and tinned doors were to be kept closed and there were to be no gaslights in the idle mill except where repairs were being made.

Overseers were supposed to be always in their rooms when they were "lighted up" and to turn out the lights and shut off the gas when leaving. They were responsible for seeing that wet, dirty waste not be left in the room (it might spontaneously combust), keeping steam pipes free from waste or lint, preventing belts from coming into contact with wood (where friction might start a fire), and seeing that fire-pails and barrels were kept full of water. Each room was required to have a number of full pails of water, according to the fire hazard there: for example, Card Room, 3 pails per 1,000 sq ft; Ring Spinning, 2; Mule Room, 3; Weaving, 1. Friction matches were barred from the complex, with the exception of the counting room, and lighted cigars and pipes were forbidden not only in the mills, as cigarettes were, but even in the yard.[8] The threat of fire hung over the mill like a pall. Everyone connected with the mill was aware of the possibility of plant, and jobs, disappearing, if not in a puff of smoke then in a billow. The full force, intellectual and coercive, of the mighty Lowell corporations was continually brought to bear against this danger.

The technology of Lowell was carefully organized to protect against fire, to assure maintenance, and to maximize production. When the enormous operation was shifted to new products, despite the inertia of plant, equipment, and workforce, the latter bore the brunt of the wrenching effect of those changes. Jobs became more difficult even as they were speeded up and stretched out, creating a magnified impact on the employees.

WORKING CONDITIONS

While the "powers that were" continually patrolled the mills performing inspections relative to fire danger, they took no collective note of other conditions there. Simply put, plants were expensive to replace and mutually insured, making their safety of greater significance to investors than the workers' safety. The disparity between attitudes toward buildings and toward workers led the Commonwealth of Massachusetts to inspect the mills' premises to protect the employees against too harsh an implementation of the system that equated minimum expenses with maximum profit. The state's experience with the cotton industry had shown that management was not to be trusted in matters regarding safety, sanitation, age laws, or even in fulfilling their agreement to pay fairly the low wages they offered. Rufus C. Wade, Chief of the Massachusetts District Police, Chief Inspector of Factories, Workshops, and Public Buildings, and Fire Marshal, noted twenty-six inspections exclusively for textile factories:

> To see that machinery is properly fenced, to look after the safety of the elevators, to inspect the sanitary arrangements of the mill, to see that the means of egress in case of fire are adequate, and that all young persons under twenty-one years of age employed in the mill have had or are receiving a proper elementary education, to prevent time-cribbing, and to see that the particulars clause . . . is complied with, so that every weaver may be able to check the amount of his or her wages.[9]

Young, who cited Wade, pointed out the difference between the two countries' regulation: "The 'particulars clause' which I have mentioned differs from ours in allowing the employer a margin of 5 per cent in the length of the cuts; that is, the employer may agree to pay 20 cents per cut of 50 yards, and require the weaver to weave 52 1/2 yards for 20 cents." The prospect for an extra 5 percent profit was not overlooked. Worse yet, a Boott weaver complained that despite a full day's work, his pay envelope was marked "Void." Another who produced twenty-three cuts of fifty yards each received just 57 cents. Investigation by Factory Inspector Frank C. Wasley led to payment of $1.37 to the first, $7.50 to the second.[10]

While 1868 had seen the enactment of a federal 8-hour law (and President Grant directed, the next year, that it should not be accompanied by a pay cut), textile mills such as the Boott lay far outside its effect. On the other hand, Massachusetts, as part of its efforts to rein in the industrialists' exploitation of the workers, in 1874 limited the workweek of women and children to 60 hours, and in 1893 to 58 hours. The law permitted exceptions, and they were sought for periods of weeks at a time. For example, when Card Room rearrangement cut production, thus slowing or stopping other parts of the mill, Cumnock asked Wade for permission to run the card rooms in Mills #1, #2, and #6 overtime, while assuring him that it could be done "without injury to the health of the children, young persons, and women affected thereby."[11] Wade routinely granted permission, just as did Francis for the use of water during off hours.

The Commonwealth also legislated requirements for pay by the week in 1886, an employer's liability law in 1887, elimination of fines for weavers in 1891, and posting of prices for specified work in cotton factories in 1894.[12] In keeping with the concept that rules are not made to prevent things that are not happening, in each of these cases the state may be presumed to have responded to existing behavior the legislators found unacceptable.

Workers had few tools with which to respond, although in times of full employment, at least, they used the one most available. During the expansion of the No. 5 mill in 1889, Cumnock wrote to the contractor: "Our help in the No. 5 Mill are complaining very much about the darkness caused by board partition, and some have left us on that a/c. We must have the windows put in *at once* in the two sections now completed and we will remove the partition" [emphasis in original].[13] They could and did vote with their feet, but the irregular pattern of employment in the industry limited the utility of such a tactic. The episode indicated, on the other hand, the possibility of working on and in the buildings simultaneously, making it clear that the stoppages in the Civil War stemmed from a desire to profit from inventory, not a necessary side-effect of modifying the plants.

The state's determination to police the safety of the textile mills offers strong, if indirect, evidence of dangerous conditions. The agent's Letterbook from the Boott presents an opportunity for assessment of the situation for the period 1888–91. Cumnock, or his assistant, filled out accident forms for all lost-time accidents.[14] One form went to the

Chief of District Police Wade, the other to the insurance company. They contained essentially the same information: name, age, job, and room of the employee, nature and extent of the accident, assignment of blame and liability, and resolution of the incident. The statistics demonstrate clearly that the mill was a dangerous place to work. Two people died and near-misses occurred. Assistant Yard Master Melville Crawford was only slightly injured after a string on his hip boots caught on the shaft of a water pump during the installation of a new Swain turbine in the #4 wheelpit. However, it was a "narrow escape from death" when "Mr. Swain threw off belt," halting the pump. "If he had gone around shaft it would have knocked his brains out against wall."[15] The instant response of the turbine's inventor saved Crawford's life. Five major injuries, such as those requiring amputation, occurred, as well as a dozen cases of loss of fingers or parts thereof. Seventy-one serious accidents affected employees of the mill (in addition to the accidents and a death of Locks and Canals workers at the Boott).

The most common incidents involved cleaning machinery while it was in operation, feeding by hand rather than with implements, and attempting to remove an obstruction or replace a drive belt without stopping the machinery. While these incidents appear to have involved young and new employees more than would be statistically indicated, workers of all sorts and at all levels appeared with regularity.

The forms reveal management thinking as well as safety information. They provided assessment of blame, and, not surprisingly, seldom found the mill to have been responsible or even that the event was truly an "accident." Employees received blame in nearly every instance. Most often they were said to have been careless, or to have violated rules or policies such as those against cleaning machinery while in operation or feeding by hand. Sometimes they did things they "had been instructed not to," or were hurt by something which "work didn't require him to meddle with it in any manner."[16] When John Cox fell down an elevator shaft (stuck open), he had no business there, had "no right to step on hatch." The insurance company endorsed this approach, agreeing that "no liability could attach to the Boott Cotton Mills, the only question being whether the hatch ought to have been closed." Although the insurance agent was willing to pay for doctor's bills, "if Cox threatens to make trouble we should prefer to have your agent do something moderate for him, if a suit can be avoided in that way."[17] Payment to avoid suits appeared to be good business. The broken-fingered shear-tender had been told "not to put his hands near the shears," and the broken-nosed lap hand warned "to be careful not to run his rack against anything."[18] Similarly, in even less clear-cut situations the agent found it easy to absolve the mill. When Ann Lawrenson slipped on a washed floor and cut her head, "She did not exercise proper care in passing over wet floor." Ellen Barett, 48, fell through an open hole behind her spinning frame where the floor was being repaired, "as she knew." The carpenter "shouted to her, but being quite deaf, she disregarded warning." Workers were so inherently culpable that they "disregarded" that which they couldn't hear; "careless" was the verdict. Similarly, when James Barker, 25, stepped on a nail in a piece of old flooring in the cloth room where he worked: "We do not consider this Co. to blame, as Barker knew flooring was being repaired, and that old floor was being thrown out as removed, with the nails in it." Worker beware.[19]

Even when management recognized a hazard, it found itself blameless, as when James McQuillan, 13, a sweeper in spooling, caught his hand in a chain on a spooler, breaking one finger and cutting off the end of a second. "Carelessness," not the mill, was to blame, for the machine was "just the same as Lowell Machine Shops are building for all their customers." The only change for the better protection of employees would be to have a "casing" for the chain, "which is not applied at present time." Knowing a remedy did not require applying it, and behaving as everyone else did was adequate defense. Even when Maurice Nelson, about 26, fell through a temporary railing at an elevator shaft while moving one of the giant spinning mules, he "must have been dizzy and lost his balance." The company's guiltlessness went far toward explaining the need for state regulation.[20]

Never in these accounts did the pressures of speed and production associated with piecework appear. These workers had to hurry if they were to make any money, despite the well-recognized risks inherent in the situation. Management could warn against dangers and make rules about procedures, but as long as it demanded high speed to attain even low wages it could not truly escape responsibility. If workers cleaned frames while they were running, or if Rosa Swift lost half a finger when Kate Kelley started one of the drawing frames they shared before Swift finished cleaning it, pressure for speed has to be acknowledged as contributory, at the least. For management, however, these "run of the mill" accidents either violated dictated procedures or evidenced lack of sufficient care regarding known hazards.[21]

Even malfunctions did not lead to shared blame. Thomas Wyman, 14, a sweeper trucking waste, stepped on the elevator hatch cover to see where the elevator stood; a bobbin in the catch had prevented half the cover from closing and he fell to the floor below. Then Thomas McGrale, 15, "ran to the hatch to see where Wyman had gone and fell through the same opening." Despite the open hatch, *only* carelessness was to blame. Cumnock indicated he was not altogether unaware of the nature of this process when he described an accident involving weaver

Mary Sullivan: "She thought she saw a rat," fell, and cut her finger when it caught in a cam-shaft gear. "The rat was wholly to blame," he noted, as if the absurdity of the standard refrain was evident even to him.[22]

Exceptions to the rule indicated two types of discrimination in what otherwise appeared to be unvarying policy. When Joseph Neaphen, 19, got his finger caught in a gear while his hand rested on a mule's hank clock and he lost a joint, the mill did "not do anything for him pecuniarily" for he was "grossly violating orders." He had "express orders not to handle or meddle with the clock."[23] Hank clocks recorded the production according to which mulespinners were paid. The report implied that Neaphen was attempting dishonestly to alter the clock. Such a concept, of course, placed those whose injuries were blamed on carelessness or violation of other rules in a separate category, recognizing, perhaps, the pressures under which they worked and the difference between their situation and one in which an employee broke the rules for advantage only to self. However, employees in each case broke rules to increase their pay. The management therefore recognized a difference between attempting to cheat the corporation and attempting to maximize production and income through speed, tacitly approving of evading rules which otherwise slowed production.

In case of accident, the Boott seemed to be of two minds: "It has always been our custom to do this [pay the doctor] for employees who get maimed in our mills, and this case also is a special one, as Ormandy is a very poor man and has a numerous family depending upon his labor." This carpenter, who sawed his finger, clearly met their discretionary criteria for the support which they described as customary. On the other hand, the case of Mary Mclullough, 30, a speeder tender who lost a finger joint in the gears of her machine, was complicated by the fact that her "acknowledged" carelessness in getting into the "properly protected" gears seemed to absolve the mill. Still, "although the Boott Cotton Mills are not at all responsible for this accident, possibly it would be better to observe the custom in such cases, and settle for the medical attendances." The implication was that while treatment was promised only if the mill were at fault, it was often provided in other cases according to *custom,* not responsibility. A desire to settle, to close accounts against further demands, often led to assumption of small expenses.[24]

Circumstances produced other responses. In one, fifteen-year-old shear tender Patrick Quinn received payment in lieu of wages while recovering from being hit in the head by a broken belt, a chance event, despite the claim of "no fault" and "no defect." Frank Trowbridge, a skilled gas and steam piper, was not chastised when he reached into a machine and was cut. Nor was James Ridings, a temporary second hand in spinning, blamed when he slipped in an alley between the frames and deeply punctured his side when he fell on an oil can. Unlike the case of Ann Lawrenson, who did not exhibit proper care while crossing a wet floor, a skilled man was blameless when he fell on a dry one. When James A. McMullan, 28, a carpenter working to level-up a card, was hurt after his assistant removed a gear cover to facilitate the work, blame was not attributed to him: his overalls were caught by the gear, drawing him in, and "his private parts were cut worst and required 13 stitches to draw the parts together. . . . He is a good, faithful—one of our best—and was working in the interest of this co. and we feel that it would be only fair to continue his wages ($1.70 per day) during the time of disability." When the injury affected a skilled employee, the response went much further than in other cases. The opportunity for discretion was the key.[25]

Only in the case of death did the mill override its own distinctions: "As is customary with us, where an employee dies in poor circumstances financially, we have agreed to assume a modest expense for the funeral, which our co. will pay if you think the Ins. Co., under the circumstances, ought not to pay it." Anson Miller, 39, watchman, had died from a fall while taking an unauthorized short-cut through the coal pocket. The "circumstances" were his violation of rules, a violation his brother agreed he had committed. The insurance company might worry about appearing to accept liability, therefore the mill could make the payments if required. As it happened, the insurance company paid $52.05 to the widow, "which she needed for immediate use," and the Boott assumed the $77 funeral expense. In a time before workman's compensation, such payments could have little effect on the condition of the family left without husband or father.[26]

In every case, the danger of liability was carefully avoided. They regularly sought and recorded the agreement of a relative that the injured had been careless or the mill blameless. "We have been remarkably fortunate in not having a single law suit. . . . [We] have endeavored to make each case a personal one." When John Moran fell in a hole in a railroad car from which the flooring had been removed, although he had been warned and was considered at fault, "there might be a chance of his bringing suit against us if we didn't care for him." When Ann Lawrenson "seem[ed] in no hurry to pay" the $14 bill resulting from her accident and the stitches and doctor's visits following it, the mill decided paying it would be cheaper than risking a lawsuit. Other workers made claims that were paid when it appeared they would "settle for a small sum." The corporation resisted major claims and those involving an attorney.

In a number of cases, it made payments for medical care in cases with liability unclear, particularly if the payment would enable the injured to return home, and "thus save board bills, doctors bills, and possibly *law* case." An extra $10 to get the person home, away from Lowell, was money "judiciously spent." Small sums of money, token payments, and other strategies to avoid expense represent the result of earlier changes in the concept of corporate responsibility for its members.[27]

The inspections and regulation by the Commonwealth indicated mistrust of the operators of textile mills in general, but the Boott came under scrutiny individually, as well. Parker raised questions about the attitudes revealed by conditions in the mill. He noted the unpleasant conditions in the Boott, then commented on the type of employees who would come to work in these dark and gloomy buildings:

> Operatives are not attracted to such rooms, seek employment there only because they cannot get it elsewhere, and are constantly discouraged in an honest effort to do good work. The best of machinery cannot be arranged or operated to the fullest advantage in such mills. . . . Excessive vibration of floors, and rocking of whole buildings is a common complaint throughout the yard, and it is unnecessary to say that machinery, however well made, set up and cared for, is soon injured more or less by such conditions.[28]

These conditions, the results of policies R. G. Dun decried, had well-known repercussions for operation, as well as for employees. For example, in 1875 F. A. Leigh gave a presentation on "Repairs in Cotton Mills" to the New England Cotton Manufacturers Association. He noted: "Most of the breakage [of machinery] will be caused by neglect, and nearly all could be avoided by cleanliness, proper oiling, and keeping machinery level, so that the shafts, rollers, and bearings are not bent down by untrue floors, caused generally by the settling or overweighting of buildings."[29] Leigh indicated significant consequences of neglect which affected both plant and employees. Such conditions, particularly in conjunction with high dividends, indicated lack of concern with perpetuating either aspect of the operation. The company responded to injured workers according to its determination of whether or not they were "working in the interest of this company" (such as hurrying production), but at the same time it treated that interest as short-term in its own conduct.

WAGES

The record of wages paid to Boott workers offers a view of daily earnings, indirectly of annual income, and of management's idea of the relative importance or difficulty of replacing certain skills.[30] A payroll entitled "Five Weeks Ending Saturday Evening, November 7th 1874" gave "Names, Time, Rate [and] Amount" for the workers in one of the card rooms. (The payroll also gave hours worked each day during the period for each worker.) The standard 60-hour week consisted of five days of 10 1/4 hours and 8 3/4 hours on Saturday.

First, second, and third hands, the room supervisors on corresponding shifts, worked in the vicinity of 300 hours at rates ranging from 17.5 to 28.2 cents per hour and accumulated $52–$72 for the five-week period. Card grinders, men who sharpened the card clothing, earned 15.5 cents an hour and worked about 290 hours, for income of $43–$46. Men who oiled the machinery and ran the elevators earned 12.7 cents, worked nearly 300 hours, and received $38. Most of the twenty-six Card Room employees fell under the heading "Roving and Strippers," those who tended the roving frames and those who stripped, or cleaned, the card clothing periodically as dirt and waste clogged the wire. While the hourly rate varied only from 12.5 to 12.2 cents, time worked ranged from a day and a half for some female roving hands to 300 hours for four men. Pay for the women was listed as roughly $20 ($1/day), two-thirds that of the men. Last were four women sweepers earning 7.5–6.0 cents per hour, generally for 24 days, for a total of $14–$18. The fact that workers often put in less than a full week reflected the common practice of running the mills only as market conditions demanded; many workers therefore were employed just nine months per year, a factor to be borne in mind when considering the already low hourly or piece rates.

This payroll presented other interesting information, as well. A perusal of the names in each category not only injects working people into the history, unusual in itself at this date, but also reveals the shifting ethnic composition of the group, with the English-named males concentrated in the top jobs, and increasing numbers of the presumably more recent immigrants, with Irish and occasionally French surnames, and increasing numbers of women in the lower-paid positions:

1st, 2nd, and 3rd Hands	*Grinders*
Ruel J. Walker	James Casey
Alonzo M. Bartlett	Martin Sexton
Solon R. Henderson	John Kenyon
Frank Coverly	Patrick Riley
Patrick Callahan	J. Irving Forbes

Oilers and Elevators
Hugh Golden
James F. Kenyon
Hiram Gilman
Anson S. Miller

Roving and Strippers (men)
Michael Riley
John Magee
Thomas McCarty
Lewis LaCross
Michal Lalley
Charles Ragan
Richard Gewan

Roving and Strippers (women)
Maggie Lynch
Elizabeth Kennedy
Mary McVey
Sarah Boyle
Sarah Leach

Roving and Strippers (women)
Alzina Lord
Mary Sullivan
Mary Ells
Harriet Jay
Elizabeth Adams
Julia Roach
Rose Boyle
Kate Dorothys
Bridget Hart
Emma Sheldon
Ann Ward
Bridget Costello

Sweepers
Mary Fitzsimonds
Margaret Murphy
Mary Keefe
Mary Sullivan[31]

Further payrolls indicate three pay cuts hit the Boott during 1874 and 1875. Jobs move between hourly and piecework, "lap boys" and "alley hands" appear, and a slight diminution in hours followed Massachusetts 10-hour law for women and children.

Salaries in the spinning departments followed a similar course. Wages declined in 1873, 1875, and twice in 1876. Contrasts between rates in a Mule Room are striking: back boys got half what doffers earned, and doffers made half or less what third hands earned. The mulespinners were paid in mills per pound according to the number, or weight, of the yarn they were making on the two mules (thus 1,800 spindles, for example) they tended. Finer yarn required the same amount of time to spin and broke more often, so it commanded a higher price per pound. In 1876 mulespinners earned $1.48–$1.52 per day on average, placing them just below the third hands. The staff of the Mule Room included one second and one third hand, two doffers, and seven back boys, in addition to six mulespinners for the twelve mules.

Pay declined similarly in weaving. These pay-cuts were not restricted to the Boott, or even to Lowell, but reflected the declining economy nationally. Their significance lay in the degree to which the owners remained able to pass along the effects of such economic difficulties to the workers. After 1880, weaving prices climbed steadily, although it would be some time before they equalled the level before the 1876 cuts. While the change in rates is clear, total daily earnings cannot be

TABLE 1.
Wage Distribution, 1889, by Income and Gender

	Males	Females	Total	%
Under $5	181	715	896	43
$5–$6	148	309	457	22
$6–$7	133	172	305	15
$7–$8	94	103	197	10
$8–$9	54	61	115	6
$9–$10	31	11	42	
$10–$12	26	1	27	2
$12–$15	6	0	6	
$15–$20	16	0	16	
Totals	689	1,372	2,061	98

Source: Cumnock to Clark, *Letterbook,* p. 429 (except percentages).

determined, since the weavers' piece-work wages were susceptible to influence from changes in loom speeds, the machines' downtime, and the amount of repairs the weaver had to make because of faulty yarn. Furthermore, the disparity between prices for first and second quality goods was often about 50 percent. Since the faults could largely be avoided when the cloth was cut for use, these "fines" were in effect more penalties than reflections of damage.[32]

As fancy weaving on dobby looms appeared in the mills, piece rates skyrocketed. In fancy goods, faults did lower the value of the material more, providing part of an explanation for extreme differences between rates for first and second quality. Since they required much more supervision from the weaver, fewer looms could be tended, so prices per unit woven had to be much higher. When instituting production of a new material, setting prices was speculative, often leading to a sudden early revision. As the mill's entire production process became geared to the new product, problems diminished and weavers produced more cloth and rates gradually fell, theoretically without a loss of total pay.

A letter from Cumnock to Clarke in 1889 (in reply to a Labor Bureau request) reveals wage distribution according to income and sex (see table 1). The preponderance of women in the lowest-paying jobs was as obvious as their absence from the higher-paying ones, but the number of men now occupying such jobs represented a noteworthy change from earlier times, the days of Yankee women in nearly all the lower-paying jobs. Male dominance of the best-paying jobs, on the other hand, stood as evidence of the continuation of the basic system as originally established. The figures support Lahne's claim that in the 1880s males had begun to displace women in the textile workforce. In addition to the wage differentials noted previously, daily bonuses of $.75 to $1.25 (more than the pay of over half the workers) paid to fifteen overseers increased the disparity of income.[33]

Average daily earnings in the New England cotton industry in general declined through 1879, recovered somewhat to mid-1883, declined again through 1884 and 1885, before rising through the end of 1890. After 1893 the slide resumed, and earnings declined steadily from 1895 to 1899. The national boom and bust economy was continually revealed in textile wages.

Indicative of the pervasive nature of the problem of intermittent employment, in 1885 in ten Massachusetts textile cities "the frequency of unemployment ranged from 14.2 percent to 62.8 percent, with a mean duration of 3.7 to 5 months." Similar figures were recorded ten years later. Even when wages rose during the period the effect was undermined by unemployment and outweighed by increases in the cost of living.[34]

Both before and for forty-five years after the Civil War, cotton mill operatives' real earnings declined relative to those in other industries.[35] According to Fidelia O. Brown, "The depression of 1893 cut back work to half-time at all the mills for several years; many operatives who formerly would have protested now felt fortunate to have even starvation level pay."[36] When the entire economy was in trouble, the workers had little hope for improvement.

Average pay per worker per year at the Boott had risen slightly by 1902, to $7.8064 per week, or $398.13 for a year of fifty-one weeks. Average 1900 Massachusetts cotton mill employee income was $351.06, placing the Boott slightly above average in theory, as would be expected for a mill making increasingly fine products, but according to Robert G. Layer, Lowell workers averaged just under thirty-eight weeks of work in 1902, which would have meant less than $300 earned.[37]

In addition to paying workers poorly, the Boott used them inefficiently by 1904. Parker calculated that Boott operations were wasteful and over-expensive, utilizing 11.42 hands per thousand spindles, despite its complement of Draper looms, whereas the work should have required but 8 3/4 to 9 hands per thousand spindles.[38] The wages and conditions impaired both effort and effect.

Chapter Five

Labor and Management

Workforce

The period from the Civil War to 1905 saw a number of changes related to gender and ethnicity in the composition of the Boott labor force. In the Massachusetts cotton industry in general, the proportion of men increased steadily, from 32 percent in 1870 to 47 percent in 1905. Women workers' share declined, from 51 percent to 40 percent, as did children's, from 17 percent to 13 percent. Total employment increased, however, from 135,000 to 310,000.[1] The number of people employed at the Boott increased from 1,310 in 1870 to 2,003 by 1890, with males increasing from 22 percent to 34 percent and females declining from 88 percent to 66 percent; less than 3 percent of the Boott employees were children under sixteen (55 boys and 15 girls).[2] Several factors led to this shift. According to Copeland, males came not because of legislation, but due to the introduction of larger, faster machinery.[3] That explanation offers little help, however. As has been seen, the Boott adopted little that was significantly different in terms of technology during this time, and the most important change, to automatic looms, did not lead to male labor. While Dublin cites Herbert Lahne as evidence for the assertion that "only in the 1880s did management seek to displace male operatives with females," Lahne's statement is misleading in the Boott's instance.[4] Displacement could only occur as production was shifted from mules to ring frames tended by women. In fact, the Boott had increased its proportion of mule spindles, but only enough for the most marginal effect on such statistics. Some areas of traditionally male labor came to demand more staffing; for example, production in picking and carding rose to keep in step with increased output by the mill, but not disproportionately. Elsewhere, demands for strength rose only occasionally. The cost of male labor, however, fell with the rising tide of immigration. Actually, the mills had never demonstrated any opposition to male labor. Instead, they had sought, and continued to seek, *cheap* labor. The change was produced by demographics.

In 1878, well over half the 10,000 textile factory operatives in Lowell were born in the United States (a meaningful statistic but one which does not take note of the numbers of first-generation workers and may seriously underrepresent the Irish). Irish-born came next, with about 2,000 women and 1,000 men, followed by Canadians, with roughly 950 women and 320 men. Immigrant females predominated. English men *outnumbered* English women 400 to 300, reflecting the desire for experienced male labor to oversee the operation and to fill skilled positions. Englishmen, far more likely to have textile training than their Irish or French-Canadian counterparts, were hired in large numbers for those jobs in which their experience was as yet irreplaceable. English (Yankee) women had left as conditions declined, and they had been replaced by immigrants.[5]

Practice at the Boott reflected these trends in immigration. Cumnock wrote to the treasurer in 1889 regarding the requirement that signs be posted regarding insurance relative to employment: "We have in our employ a number of French people, and have no cards printed in that language. Please send us some to oblige."[6] Cumnock's request indicated a new presence and his awareness of it, and it reinforces our belief that workers of a given nationality worked together: if they couldn't read the English signs, they would have gained little from instruction by English co-workers; as had happened earlier, they were trained by friends and relatives.

The nature of the work experience was affected not only by the technological and ethnic changes noted, but also by the comparative lack of mobility of an immigrant family. Agents at a meeting of the New England Cotton Manufacturers Association quoted and objected to a description by George E. McNeill, a Deputy State Constable, of a mule-spinner's account of his workday. He described long hours at the mill for himself and his wife and ten-year-old daughter, who together tended ten looms. Work, food preparation, eating, and travel devoured 77 1/2 hours each week. Cumnock responded: "I do not know of any city in the State of Massachusetts where *all* the operatives live in this way" [emphasis added].

McNeill contrasted the situation to a past one in which workers, perhaps with less political freedom, were more nearly socially equal to their bosses and had more open futures: "Then, the spinner might leave, and become carpenter, teamster, blacksmith or grocer. Now, we have that dreadful thing, a fixed factory population,—not fixed as to locality, but as to handicraft. A hand!! A hind!! An operative!!" In his eyes, the transient labor of the Yankee era was a thing of the past, a time of greater economic alternatives. He saw a new type of cotton mill operative:

> Short of stature, mostly without beard, narrow chested; somewhat stooped; a walk not like the sailor's, but equally characteristic; not muscular, but tough; flesh with a tinge as though often greased; cheeks thin, eyes sharp,—a man pretty quick to observe, and quick to act, impulsive and generous, with a good deal of inward rebellion and outward submission. . . . Except to obey, the agent and overseer are nothing to him.[7]

McNeill objected to the new order and all that had been lost. The worker he cited lived a life of deprivation without hope of escape, without time for a life away from work (let alone for writing articles or attending lectures). Economically subjugated, socially estranged, physically blighted, this person saw neither a connection to the economy in which he or she labored nor mutual interest with those for whom they worked. The agents, however, were appalled at this characterization of the effects of their operations. They denied McNeill's charge and objected to its emerging from an employee of the state. Plainly they held themselves apart from any actions which could produce such a situation. Former workers, they did not see themselves as exploiters, but neither did they consider factory work a suitable *fixed* occupation (just as Lowell's founders and its early defenders had not).

MANAGEMENT OVERSIGHT

Work in the cotton mills in this period was not attractive. The work was monotonous, dirty, hot, and hard. Pay was low and pay-cuts routine. Workers in the Boott were handicapped by old and insubstantial buildings badly laid out for their current use. The machinery varied from state-of-the-art to archaic. Yet management also suffered from some of these problems, and remedy of these many ills lay beyond the reach of those who worked in Lowell.

According to Lubar, as agents and even, at times, treasurers came to live in Lowell, they became part of the local community and began to value steady, smooth operation of the mills without cuts in pay or hours in order to preserve their relationship with those beneath them. Cumnock's football outing certainly indicated numerous local connections, but the Boott's treasurers remained in Boston and continued to desire operation which quickly reflected market conditions.

With regard to much of what went on in Lowell, the agent was free to exercise his power as he saw fit, particularly if his acts were not costly to the stockholders. Cumnock clearly ran the Boott according to his own set of rules. When Julia Buckley reported late to work, she found her place given to another because she hadn't sent word and was supposed not to be coming. By implication, she knew the rule. Similarly, Ms. Degnan could not receive her wages until payday, because she had not been discharged but "remained out without notice." Employees could be "dishonorably discharged," as were Deba Smith, Tillie McClabe, and Henry Ashton, who "sold his pay in your No. 6 Weaveroom Apr. 15th, and again yesterday, in same room. How did you happen to employ him after he sold his pay?" Apparently, one could get money (presumably discounted) before payday in this manner. The office found it a nuisance to deal with those who "bought peoples pay" and therefore ruled against the practice. Interestingly, not only could the overseer overlook the rule, but the hated broker apparently could enter the mill and conduct business. Control appears to have been not always so complete as it would seem from management accounts.[8]

Yet from their point of view, managers had a direct interest in their employees—their conduct and their welfare both at work and when "off-duty." This interest offered the classic opportunity for the two-edged sword of paternalism, the treatment which "guarded" welfare by dictating conduct. At the same time, the concomitant taking of responsibility for the governed, the essence of true, familial paternalism (or maternalism) never appeared.

Management demonstrated concern for employees on several levels. When the Central Street Bridge burned, management saw to it that workers in Centralville were ferried to work. In response to a letter from Sarah Norcross, in Augusta, Maine, Cumnock wrote, "Am glad to hear from you to know that your health has improved once you went east," and he sent her pay of $25, plus interest. When the state became sufficiently mistrustful of the textile mills regarding their effects on child employees to require school attendance and literacy for those under sixteen, most people are said to have ignored the rules. Inquiry to the Boott asking the reason for the non-attendance of a boy at Green Evening School brought the response that "his card shows a regular attendance" for the past two months at the Moody School. Knowledge of the workers' lives appeared in less formal ways, as well. When Johanna Livingston wrote from Marietta, Pennsylvania, inquiring for Harry Livingston, the office responded that he arrived in Lowell with Foopaugh's Circus, worked in a couple rooms of the Boott without "taking much interest in the work" and was discharged, whereupon he left town with Burke's Circus.[9] There was an informality operating here regarding minor matters, a personalism not generally recorded. An interest in the employees as people appeared, whether approving or not. It also implies that Cumnock's position did not lead him to forget or deny a connection with the workers. It suggested to employees a caring oversight and interest, which it was hoped the worker would return through a belief in a mutuality of purpose, which they could indicate through dedication to their jobs.[10]

The nature, and limits, of this interest in the workers appeared in letters regarding admission of employees to the Lowell Hospital for reasons other than work-related injuries. They indicated circumstances of illness, such as those of the consumptive Maggie Bell, but willingness to provide her with the necessary admission card turned on the fact that another employee stood ready to be responsible for her costs. Admission required a card, and for that the mill required evidence of fiscal ability. Again, knowledge and even interest, without responsibility. The possible contribution of factory work to her condition was implicitly rejected by reference to her "consumptive family."[11] Such superficial interest made it equally easy for management to inquire, regarding a prospective overseer, of his nationality, or to attempt to protect employees from harm: "We desire to call your attention to the fact that crowds of loafers frequent the corner of French and John and French and Paige Sts., and offer insult to females passing by them. Many complaints have been made to us of this, and we are told that these fellows make a practice of enticing such young girls as they can to disreputable places." In the first instance, they sought information irrelevant to job performance, while in the latter they sought police protection for those who may well have desired it. Their power and interest allowed them to decide what was important.[12]

Cumnock's attitude toward the boardinghouses, their residents, and keepers made most obvious the nature and purposes of his interest. Company-owned housing still held one-quarter of the Boott employees in 1888–91. Housing and/or board were available at fixed rates, with costs withheld from the employees' pay and kept for the housekeeper. Set policies governed the conduct of both residents and keepers: "We are informed that Maggie Downing has left Mr. Dearborn [overseer] without notice, and has gone to work for the Merk Company and is still boarding with you. Miss Downing must either return to work for our company, or get board elsewhere." Keepers had to keep the names of all their customers in their book, so the company could know who was there, and only Boott workers could be accommodated without the agent's permission. Non-employees routinely used the houses, but the rules enabled the company to remove boarders if they left the Boott or broke rules of behavior, or if it wanted to provide the space to an employee. "Edward Hazard sold his pay and left our employ." After breaking two rules, he not only stayed on at the boardinghouse, but "he spends his time loafing about street corners." Mrs. Bixby must dismiss him. Continued provision of housing allowed the agent, if he wished, to extend his power into the lives of the workers while in the community. Further, it could be used to attract or accommodate an employee, again extending the agent's power.[13]

Not that maintaining the system was painless for the agent, who had to deal with the usual problems associated with being a landlord. Horses knocked over a hitching-post, a grocer wanted to hold an auction at the residence of the boardinghouse keeper he had attached, and keepers didn't always pay their water bills (for water-closets) on time, even when one "promised to send money by your little girl to settle up, but I haven't seen her." More significantly, the residents of one tenement were "throwing swill, tea grounds etc. into the cesspool," stopping the drain, rather than using swill buckets, and threw out a maggoty cat in the ash can: "You must be more cleanly, and not have as much filth about your premises, as it will breed disease, and *can't be allowed.*" Other boarders threw "filth and debris" on the tops of sheds, families brought complaints and counter-complaints against one another and their children, and a soil pipe burst in a privy vault, requiring a repairman. Problems of these sort were common and obviously a considerable nuisance. The continuation of the housing indicated that the agent found it of sufficient value to offset its inconvenience.[14]

Eviction remained Cumnock's ultimate rebuke. He had one worker ousted "for having used language in the street in front of your premises unbecoming a gentleman." Standards were based on Cumnock's beliefs, not work-related issues. By far the most common cause for dismissal was drinking:

> We are informed that Annie Driscoll—who has been working for our Mr. Dearborn—has been drunk at your house all this week.—You probably know that this is contrary to your orders from the agent and that *all cases of drunkenness must be reported at Counting room.* You will *at once* dismiss Driscoll from your house and *look out for any further violations of our rules.* [emphasis original]

Drinking led to loss of both home and work, and the housekeepers, as well as management, were expected to guard against and report it. David O'Hara lost his place and got his tenement keeper warned because he was not only drunk, but "brought bottles of liquor into the house." Annie Pierce left work "sick" but then "She went off riding with a fellow & another girl, and that he [Lawley, her overseer] saw them drunk & disorderly on the streets," as he had on another occasion. Subsequently, she went to work for the Merrimack before returning to a different room at the Boott under an assumed name. Such behavior could not be tolerated, of course, and the office notified her new overseer, and her boardinghouse keeper, to dismiss her, "As soon as her place can be filled." Temporary moral contamination of the Boott was preferable to interfering with production.[15]

Enforcement was less than absolute, as were its results. One fired employee was found to be still boarding on the Boott six weeks later. However, it often served its purpose; housekeepers regularly dismissed and reported residents for drinking. They subsequently lost their jobs. Failure to report, on the part of the housekeepers, brought reprimand and even dismissal. Mr. Hutchins, who let Annie (McGuiness) Pierce stay at his house, later heard again from the office:

> Report also comes to us that you have a great deal of drunkenness in your house, and don't report it to Counting Room. You must be careful to run your house a little more strictly, and not allow drunkenness in it, nor any profanity or disturbance and report at our office all cases of this kind which may occur, that we may purify our corporation, by discharging the offenders.

Other keepers were warned or discharged for tolerating noise or "keeping a noisy and disorderly" house.[16]

Despite the inherent difficulties of landlordism, Cumnock not only continued the burden but extended the chore by using his position as an opportunity to "purify the corporation." The concept of a moral police to govern workers outside the mill gates lived on. Regulation lacked the excuse of being work-related, just as it could no longer be claimed that strict order was needed in order to attract new workers. At this point regulation existed with no more justification than the agent's desire to enforce his will, his concept of purity. In fact, rather than proposing to attract more workers, these efforts aimed to remake the habits and beliefs of the plentiful arrivals. On the other hand, they represented an aspect of Cumnock's efforts to create distance between his concept of textile workers and that presented by McNeill.

The Boott had a more formal relationship to the city of Lowell, of course. Some changes had occurred in the city since the early days of its total domination by the corporations. Disputes over valuation of the mill and its property for tax purposes were not uncommon and could involve several thousand dollars of indebtedness. The Boott remained self-interested, as when it opposed a new sewer on the grounds that it emptied its own waste into the canal. Community interest remained separate from that of the corporation. On the other hand, relations were cordial enough to serve mutual interests. The Boott routinely requested and was granted the appointment of one of its employees as a special police officer, without cost to the city. The need for and use of this authority remain unclear. Further, the school department was granted permission to erect a temporary school building on Boott land, without charge, for two years, as long as it would be left there and be suitable for a storehouse afterwards. More routinely, the company asked police to arrest snowball-throwers on Brown Street whose harrassment of female employees at noon had brought complaint, and to stop children from playing ball in the mill's lumber yard, for they had "broken considerable glass." They also requested a fire alarm box to better protect their tenements. While disputes between the corporation and the city arose occasionally, the corporation remained a prominent taxpayer that acted in its own best interest and felt free to request services from the local authorities.

The true nature and extent of the Lowell corporations' power came from and can be seen in their ability to act jointly, an ability they cultivated. In 1880 they founded the Arkwright Club, made up of treasurers of textile mills across New England, "to promote good understanding and united action upon affairs of general interest" such as cotton buying, tariffs, and other legislation, as well as to pursue general interests in other areas, including wages and political issues. The club provided a convenient mechanism for cooperation. Through it the mills regulated their output, followed labor legislation in the State House, and generally pursued the interests of their "community."[17] During one six-month period the Boott shared expenses in opposing a fortnightly payment to operatives bill, a bankruptcy bill, and several other legislative matters. The Boott also maintained a membership in the Associated Textile Industries (ATI), which provided a mutual strike insurance for the various companies whereby they shared the cost of strikes affecting any one of them. In 1888 a letter from the ATI announced an assessment on members "to pay for aid on account of strikes during the year ending April 13, 1888." The Boott owed $133.47 of the $5,994.23 total, based on its payroll.[18]

The corporations acted jointly in other ways, as well. They shared information on the subject of rates paid for various jobs and the income they expected those rates to produce. Agents cooperated in collecting debts from workers who changed mills, and they divided the cost of hiring a private detective to spy on labor. Cooperation among agents also reached outside the community to serve industrywide interests. In 1890, for example, Cumnock wrote to E. M. Shaw, agent of the Nashua Manufacturing Company, to deny a newspaper report that the Boott employed striking Nashua workers: "We shall endeavor to stand by you so far as possible." Overseers had been instructed not to hire the strikers and to "bounce" any hired inadvertently. With the New England Cotton Manufacturers' Association established to exchange technical knowledge, and the various organizations and informal mechanisms, the original companies still wielded overwhelming clout in the mills, the community, and the region.[19]

LABOR'S RESPONSE

Despite their preponderance in numbers, the workers' response to management was fragmented. Direct evidence of behavior and attitudes in and around the workplace remains sketchy, but occasionally a dramatic event exploded the barrier of statistical anonymity and revealed the lives and passions which of course filled the vacuum that appears in histories. Cumnock described one such moment in a letter (quoted in its entirety) to Treasurer Lowell in 1878:

> At about 3:30 this P.M. Fred W. Sproat, (section hand) and Miss Laura Hunt, (weaver) employed by us, were standing in front of No. 5 Mill discussing their love affair.—She having told him she would not keep his company longer,—when he deliberately shot her twice in the head, and then shot himself. Miss Hunt is dead and there are no hopes of Mr. Sproat's recovery.

Such a tragic event reveals and combats the insubstantiality of statistical accounts, as well as the aloofness associated with paternalistic treatments, both of which rob their objects of their vitality and completeness. The slightly recorded lives of the workers, within and without the factories, went on with a complexity and intensity undiminished by its general invisibility to modern readers.

Workers not only lived their own lives but also endeavored to resist their overbearing superiors in a variety of ways which occasionally come to light. Their drinking (noted above) appears to have been considerable and consistent, despite management's disapproval. (It elicited some twenty letters from the counting house during the period covered by the Letterbook, far more than any other subject related to the boardinghouses.) The drinking episodes reported often involved groups of people, and these included both mixed and single-sex gatherings. Various nationalities were represented, with the English and Irish, the longest present and potentially most acculturated to the mill mentality, predominant. Their willingness publicly to violate established rules indicated both an independence and a continuation of a culture separate from that promoted by management. It also implied a belief that further employment could be found, as seems to have been the case. Not only did discharged workers return to the Boott and also turn up at other Lowell mills under their own names, but also, in several instances, they adopted aliases to evade the effect of a dishonorable discharge. The success of such an effort clearly revealed the existence of a worker culture in opposition to management, since it is unlikely that other workers would be unaware of the presence of an old worker with a new name.

Clear evidence also survives of something which might be called familialism. The effort revealed in the accident reports to acquire agreement from another family member regarding the fault of the injured showed that very often the victim had a relative in the mill, often in the same work area, a perpetuation of an old pattern. At this time the relationship might be parental and involve disciplining of children. When fourteen-year-old Joseph Boyle ran through an elevator-way and fell to the floor below because a piece of cotton had kept half the hatch cover from closing, the mill indicated it had no blame, and that he had "already been punished by father for same stunt." Not only were families present, they interceded their discipline between that of the bosses and their children (much in the same way described in *Like a Family* for the South, later). Similarly, they passed along their knowledge, enabling workers to move up the labor hierarchy, as in the case of an elevator-operator permitting a bobbin-boy to run the elevator. A worker culture extended beyond the workplace, as well, as is indicated by the collection of $230.55 "for relief of suffering by recent flood." These workers showed that in many ways they were a cohesive group, connected by work, family, and mutual interests. They had connections which lasted over time (when Elizabeth Mullen, a card-room worker died in 1889, one of the other workers had known her for fifty-six years).[20] They maintained their culture, their family ties, and, to a degree, control over discipline and education in the mill. In fact, they even rewarded bosses at times. When William Wood, for thirty-seven years an overseer at the Boott, retired in 1877, he was given a Turkish chair, a gold-headed cane, a vase, and a black walnut case, all presented by Edward Robinson, his successor. Such a presentation of gifts from workers to a retiring supervisor suggested a lack of distance, a feeling of shared interests, that made recognition of his role appropriate. It implied a connection among all those who worked, a connection the employees recognized and, perhaps, wished to emphasize. In addition to initial efforts to create formal organizations during this period, the workers continued to bring preindustrial and extraindustrial traditions and beliefs to bear on their work.[21]

Strikes during this period were occasional and generally were restricted to the members of one of the few craft organizations in the mill. Mulespinners, a predominantly English group known for their skill and the independence that it enabled them to demonstrate, figured in occasional organized disagreements with management. These skilled males formed craft organizations early, using their skill and the importance of their production to the mill's operation in an attempt to gain a position of strength. As early as 1874 mulespinners struck the Chicopee's Dwight Manufacturing Company in opposition to a 15 percent wage-cut, and Lowell spinners sent money in support of their effort.[22] Lowell mulespinners struck in 1875 to protest wage-cuts. The owners responded with a lock-out and won. Many of the spinners left the city, and those who returned to work could only do so on the condition that they sign "yellow-dog" contracts, a promise not to join a union.[23]

Parker's report in 1904 cast an interesting light on the conflict and its relationship to technology: "Have estimated for 45,600 new Mule spindles, and specify mules for three reasons: Fine filling yarn can be made cheaper and better on mules; they take less power; and are cheaper per spindle. You may object to mules on account of the possible trouble with Spinners Unions, and this factor will probably have a bearing on your decision."[24] Despite factors favoring mules on the basis of both quality of product and costs, he anticipated reservations about their use.

The battle between mulespinners and management continued throughout the period. According to one source, management in some instances shifted production to make heavier weights on spinning frames at the same time they lowered the weight of the mules' production, maintaining cloth weight. The lighter yarn increased mulespinners' production problems and diminished their piecework earnings.[25] By 1889 the mulespinners had created the National Cotton Mule Spinners Association of America.[26] Their advancing level of organization spurred efforts to improve the quality of frame-spun yarn in an attempt to make the mules and their operators superfluous.

Despite the long hours and low pay, operatives had been unable to draw together and resist except in the cases of a few craft unions: "These unions were represented by the Lowell Textile Council which as a federation helped coordinate the activities of the craft locals. Its president was Robert Conroy of the Beamers' union, and its fifty-six members [largely building trades] represented the unions of the carders, weavers, beamers, mule spinners, loom-fixers, knappers, and knitters."[27] According to the management-oriented *American Wool and Cotton Reporter,* workers formed unions to protect the "weak from the agression of overseers," an interpretation that contained all the problems within the workforce.[28] The predominantly Irish-American unions included a few French-Canadians but virtually none of the Poles, Portuguese, and Greeks who had come to play a major role in the Lowell mills. A strike-threat by the organization in 1902 brought a counterthreat to close the mills. A local committee investigated the corporations' claim that they couldn't afford the 10-percent increase and led to the reversal of the council's demand. When the council repeated its demand in 1903, it dismissed the committee's fundings on the basis that it had been made up of shareholders. Claiming a 25-percent cost-of-living increase between 1899 and 1903, the council sought a 10-percent wage increase, comparable to that won by Fall River mill workers the previous year. The agents responded collectively, using William S. Southworth, agent of the Massachusetts Mills, as their spokesman to reject the request. When the unions threatened to strike on March 28, 1903, the mills locked everybody out. Management later admitted that the high price of cotton and large inventories of goods would have caused them to close the mills anyway, as they did other mills. The workers could not have chosen a worse time. The United Textile Workers Association, the National Mule Spinners Association, the American Federation of Labor, and the Lowell Textile Council all supported the strike and provided assistance to union members. The 1,600 nonunion strikers received nothing: "Hundreds of Canadians are returning to Canada, and many of the best operatives are seeking work elsewhere." A parade drew widespread support, and 4,000 workers marched, including a large number of women with the weaver's division. Workers lost $250,000 per week, while the corporations sold cotton at a profit to maintain their incomes. According to one worker: "You see the mill treasurers are a philanthropic crowd, who by nature and training believe they are better qualified to care for our earnings than we are. That pet belief of theirs might have remained undisturbed for years to come if they had only been just enough to allow us sufficient to live on, but they would not do it and we have shut off their income in the hopes of bringing them to a reasonable state of mind."

After nine weeks, the workers were hungry, particularly those without union assistance. The mills announced that they would reopen June 1 and would rehire anyone who returned within a week. The Greeks voted to return and the French Canadians were expected to "welcome a chance to return to work." As had long been the case, "the mill managements played one race against another, both industrially and politically," according to the *Wool and Cotton Reporter*.[29] The Boott claimed a 45-percent return on the first day. Most of those who did not belong to a craft union returned and the mills were soon operating. Many unionists held out beyond the one-week period and were never rehired; the craft unions were permanently damaged. The workers who replaced those who left, in many cases the newer immigrants, received promotions to better jobs as a result, further damaging the union cause in

Lowell in the years to come. At this time, labor demonstrated little capacity to influence the operation of the Lowell mills through organization.

Collapse of the Mills

Crises had struck the Boott in the early 1840s, the late 1850s, the mid-1860s. Each time, pay-cuts, loans, and other forms of speculation designed to protect dividends had perpetuated the corporation. In 1904, the cycle repeated, only the solution was novel.

Despite generally steady profits in the industry in the final quarter of the nineteenth century, the type of management R. G. Dun had noted at the Boott made it less than a surprise when things came to a head in 1905:

> The affairs of the old company have been complicated for some time, and the final crash came on Feb. 4, when it was announced that the company must cease operating its plant. On that date the mills were closed and it was stated that unless a complete new corporation could be organized and new capital introduced the mills would have to be sold and further operations suspended.

While other plants continued to operate successfully in Lowell and elsewhere in New England, only a month earlier the local newspaper had recognized the Boott's vulnerability: "Even the Boott company lately picked up enough orders to keep it going for the next two or three months."[30] Simultaneously, however, the paper announced that the "Boott's Directors Given Authority to Sell Mills." Technically "given" the authority by the stockholders, the directors controlled sufficient stock to make their will dominant. A. S. Covel (Treasurer), H. B. Lincoln (Clerk), Arthur T. Lyman (President), Jacob Rogers, Charles F. Ayer, Charles Lowell, Charles F. Adams II, and Arthur Lyman controlled the operation through both family and economic connections. Thirty stockholders at the fateful meeting controlled 889 of the 1,200 shares outstanding (leaving little room for the famous widows and orphans said to be dependent on such stock whenever anyone had the temerity to appear to threaten the corporations' interests). Some, such as Lowell, had roots which went back to the company's founding, while others, such as Ayer and Rogers, would play major roles in the future. The paper blamed a collapse in the price of cotton the previous spring, following a "speculative advance," in combination with a great depression in the price of goods, "an almost unparallelled abstinence from buying, especially of fine goods."[31] The difficult circumstances had already forced the Boott to close for three months before reopening in November 1904.

The dislocation was more than an incident in the roller-coaster economy of the period, according to a Boott memorandum from the 1950s: "In 1904, when Mr. Sully, of Providence and New York, son-in-law of D. M. Thompson, head of B. B. & R. Knight group of mills, cornered the cotton market and sent it up to 16.65c per pound, Boott and all of the other mills had to buy some cotton at least. . . . [and after the price dropped to 7c by January, 1905] every mill lost money." When the cotton market reacted, "the Boott Cotton Mills, instead of taking advantage of the improvement in the cotton cloth market, sold out. . . . Shareholders realized only $45 per share for stock with a par value of $1000."[32] The purposeful manipulation of the cotton market for personal benefit apparently succeeded in enriching Mr. Sully, and created months of unemployment and privation for thousands of workers. Since other companies did not fail, that incident cannot bear full responsibility for the Boott's problems, however.

The Boott had paid a dividend of 6 percent in 1900, and of 2 percent in 1902 and 1903. Yet its stock had fallen from highs well above $1,000 per share in the 1880s to roughly $250 by 1904, a price which indicated a value for the plant of only $300,000, according to the *Citizen.* The paper cited the unsettled cotton market and contrasted the Boott's crowded site, dark, tall buildings, and old-fashioned machinery with the larger buildings, modern machinery, and lower-priced labor available in the South.[33] Yet the other Lowell corporations did not fail at this time, and the Boott was not sending its money South to take advantage of conditions there. Boott management had long been aware of the problems inherent in its operation and had refused to make the investment of income in buildings and machinery which would have continued the operation's viability. In fact, as recently as 1903 they had hired Lockwood-Greene, and Company to prepare "Sketch Plans Showing Proposed Reorganization of Boott Cotton Mills, Lowell, Mass."[34]

For some time the mill's selling house, Smith, Hogg and Company, had been underwriting the mill by endorsing its notes and even paying salaries and wages. By 1905 they held $800,000 of indebtedness, balanced by a valuation of cotton and salable goods of $600,000, plus $300,000-worth of physical plant and equipment, leaving $100,000 for dispersal among the stockholders after Smith, Hogg, was satisfied.[35] Lowell corporations traditionally left the marketing end of their business to selling houses. It had often been felt that their principals, connected to corporate leadership, received large sums from commissions at little or no risk while the manufacturer of the cloth worked hard, took risks, and made money less easily or reliably. Over time, the selling houses became increasingly powerful. They not only advised mills on marketable production, they also provided credit as needed, interposing themselves between corporations and the financial centers of Boston and New York. They often owned substantial portions of the operations for which they sold. Stockholders and observers had long objected to the role of the selling houses, at worst as "parasitical intermediaries."[36] The lack of outcry from the directors when the Boott collapsed indicated the extent to which they had recognized the ultimate result of their style of management. They had been paid off with earlier dividends. When the Boott Cotton Mills had been milked to the point of collapse, owners had already received sufficient benefit from their investment. Workers were idled, without income, first for three months of hard times in 1904, then in 1905 while the company's future was debated.

Conclusion

For the Boott, the period 1871–1905 was one of expansion and change. Production became much more complex and varied, runs became shorter. These changes combined with increases in the size of the operation, the automaticity and speed of the machinery, and the relative contraction of the workforce to increase the difficulties of production manyfold for all parties. They placed the Boott in competition with factories in other regions, such as Philadelphia and Rhode Island, where such operation was typical of much smaller plants. The "tidy hierarchical system" of the East Scranton postulates for the 1900 period had passed at the Boott.[37]

Cumnock presided over the shift to a variety and order mill, as well as the change from operation as a series of semi-independent mills to a unified plant, where #6, for example, supplied filling yarns to 882 looms in mills #1, 2, 3, 6, and 7 in 1895. Operating the mills as a unit, making the new fabrics, Cumnock had to do a great deal more management and handle far more complexity than had his predecessors. However, his success did not prevent him from pointing out inherent difficulties. Pressed to produce three and four harness twills in 39" widths, he indicated that it was already done in 30", and that it was "pretty expensive fitting up twill motions, reeds, harnesses, etc." for the wider goods. His objections were not effective, however.[38]

Workers struggled to handle increased loads at faster machines making more varied, more fragile materials in ever-changing runs of smaller quantities. (When Cumnock observed in 1875 that in the time from

January to April he had more complaints regarding tender yarn than in the three years previous, we hear the faint echoes of innumerable complaints from spinners, weavers, and everyone else working with the yarn that it made their work harder, less remunerative, lower quality, and came at a time of falling pay, as well. Cumnock blamed the bad cotton crop.) Furthermore, observers recognized that the new machinery was not truly labor-saving: "Under modern industrial conditions, with automatic machinery running at higher speed, the fifty-six hours a week in the cotton mills bring greater exhaustion of body and mind than the seventy hours of earlier days." No longer did workers sew or read, but "now the operative must be alert, or else become an automaton, a part of the machinery he tends." At this time "brain-fatigue" acted to "devitalize" the worker. As David Montgomery notes, "The urgent task confronting northern textile manufacturers . . . was that of using labor more intensively—of raising productivity." More than doubling production with a virtually static labor force after 1875 certainly fulfilled this desire. At the same time, the lack of alternative employment maintained the supply of employees for these jobs. Management had long blocked the entry into Lowell of other major employers. Through their organizations, they also prevented the workers from unionizing effectively. From their point of view, only they had the capability and right to make the decisions by which the mills, the community, and the lives of the workers would be run. Conflict persisted: automaticity versus flexibility, maximum efficiency versus minimum investment, management's interest versus labor's desires, all in an environment where air full of fly endangered everyone's health and led to the derogatory term for cotton worker: "lint head."[39]

The endemic nature of such conditions, and the companies' apparent unwillingness to alter them, led to increasing inspection by the Commonwealth to protect the workers from the effects of the operation of the Lowell system. Efforts were made to police the wages, hours, safety, ages, and education of the employees.

The economy in which the Boott operated was less subject to analysis or control than production issues which were susceptible to engineers' scrutiny. If dividends were to be maintained, management had to manipulate production. As has been shown, keeping pay-rates low offered a prime opportunity to maximize the return on investment, as did avoiding reinvestment in plant and equipment. As a result, buildings and machinery did not remain current, and working conditions and wages declined.

During this period, Lowell and New England saw their comparative monopolization of the cotton industry begin to erode. While before 1880 cotton production in the South was unthreatening, after that date it mushroomed. Southern initiatives opened many mills, some indigenous, some assisted by northern investment, some branches of northern operations. Northern machine builders extended credit and even bought stock to assist southern efforts with minimal capital. By 1903 a need for an alternative to the New England Cotton Manufacturers' Association produced the American Cotton Manufacturers' Association, a primarily southern group. On the other hand, at the same time two-thirds of the cotton spindles and three-fourths of the looms still ran in New England.[40]

Thus northern and southern investment in new capacity in the new manufacturing regions came before any demonstration of labor strength or organization. In fact, the mills enjoyed new waves of unindustrialized immigrant employees during these years, an influx which helped perpetuate the mills' unorganized character for years to come.

Southern operations offered northern investors further, rather than alternate, opportunities. For example, Cumnock left the Boott to take over the Appleton, where he replaced old buildings in an effort to create ideal manufacturing conditions, a goal he did not seem to have been able to pursue while under the reign of Boott directors. At the same time that he directed this massive investment in Lowell, he initiated operations by the Appleton in South Carolina.[41] His decision was not to leave Lowell or avoid any problems there, but to develop an expanded capacity elsewhere as well, and the South offered a convenient opportunity.

For the Boott, the period was one of consistent dividends. Stock prices remained high until after 1890, and even then stood at par or better. Dividends averaged 4 percent through the turn of the century; however, the industry as a whole was paying 8 percent at that time. Dividends remained consistent in part because of the practice of paying them on a basis "closer to gross than net earnings." While the plant might deteriorate, dividends continued.[42]

Recognizing the deterioration in the Boott's condition, the selling house advanced it cash only to the extent its risk remained easily redeemable. Recognizing the operation's weakness as resulting from policies which had produced profits while limiting future viability, the directors accepted the demise without complaint.

Part Three

Reorganization and Success, 1905–1930

CHAPTER SIX

Direction of the New Boott Mills

THE RECORD of this period offers a new thoroughness of information about the mill, its operations, its workers, and their tasks. More identifiable individuals enter the story, and the activities of all the people associated with the Boott achieve a new visibility, heightened definition. New leadership takes over the operation: first Frederick A. Flather becomes treasurer, then his sons, John Rogers and Frederick, join the staff. The stories of the Boott Mills and the Flather family intertwine.

The study of management and the "scientific" analysis of production now promised that worker productivity could be increased without increased exertion, and at minimal expense. Partly because of such views, Boott management repeatedly hired consultants who examined and reported on the quality of the buildings, the machinery, the employees, and the administration of all of them. Rather than the generality of "textile work," descriptions emerge form these consultants' reports which bring the jobs and work-spaces to life, reveal their condition, help relate wages to the cost of living, and describe oversight and work. Finally, labor-management relations during this period offer insight into ways of thinking about the work, the mill, and its place in society.

REORGANIZATION

After the Boott's collapse early in 1905, and during efforts to reorganize it, the previous directors revealed their attitudes through their actions: some left the company, while others organized the move to take it over and renew operations. The Ayer family increased its involvement, and Frank E. Dunbar led the effort to purchase the mills for Jacob Rogers, his father-in-law. While workers waited for wages, investors wiped the slate clean for a fresh start.

The failure of the Boott in 1905 revealed the effects of lack of reinvestment, particularly in buildings and maintenance, over the last decades. While Cumnock's astute leadership had kept the operation going and produced new products to avoid competition with more efficient producers of staples, the steady payment of dividends had obligated profits and blocked strategies, such as those suggested by Lockwood-Greene, aimed at perpetuating the operation. The "new" owners felt that continued implementation of the old course of action (i.e., Cumnock's) could derive further profits from the Boott, its plant and employees. They didn't plan to suddenly reverse past practice, to invest in modern buildings and a new complement of machinery. Rather, they would continue to seek success through market strategies, through refinements of practices as suggested by consultants, and through the extraction of the maximum production at the lowest cost from labor. They paid debts, changed some personnel, and restructured the investment. Flather later claimed it was "not a purchase in the open market" but one made "without any reference to the value of the property." His description of the transaction suggested that the general stockholders were deprived of an asset he valued at $2.2 million through an insider trading deal. Those holding shares which recently sold for as much as $1,400 were sold out for four cents on the dollar.[1] However, a contemporary evaluation placed the value of the plant at $300,000, the amount paid for it in the reorganization. The directors who left the company plainly accepted the lower valuation and the payoff it implied. The closing and sale were common knowledge. If other parties had considered it attractive, they could have made counter proposals, as has so often happened in recent years when efforts to buy out a company were initiated. Flather inflated the value for his own purposes, to defend the manner in which the finances were handled during his tenure.

In a retrospective view at the time of the mill's final demise, the Flathers described the first "failure" and its aftermath:

> In February 1905, the cotton market reacted strongly [from the downturn caused by Sully's cotton corner]. The Boott Cotton Mills, instead of taking advantage of the improvement in the cotton cloth market, sold out for $300,000. . . . At this time, silver was high and cotton was low, and China had no end of silver. China ordered more cloth than ever before or since, and employed the mills in Lowell and many other mills in New England on drills and sheetings. The Boott Cotton Mills could easily have been sold in a "buyers' Market" in 1905 and 1906 which followed at a much higher price.[2]

Dunbar put together a group to buy the plant and equipment. Although he had hoped to raise the necessary money in Lowell and had reported assurances of support, when the time came only about one-fourth of the total capital came from local residents, the rest primarily from Boston interests. The investors authorized Dunbar to pay no more than $300,000 for the plant and assets and to organize a new company with a capital stock not to exceed $600,000, of which they claimed to have $500,000, including the purchase price. Previous stockholders would be given the first opportunity to buy remaining shares. Such an approach to renewed operation enabled the major players to keep control of the Boott consolidated. Of the $600,000 said to be invested, however, they actually raised only half, another old Lowell practice to diminish the apparent percentage of dividends. Despite Flather's claim that, "We may say that the additional $300,000 of so-called 'water' which was originally added to the Plant Account is barely enough to represent the water rights and L. & C. shares," in a discussion of *investment,* such a statement of value is irrelevant. The misrepresentation meant that New York selling house Wellington, Sears and Company's pledge of $250,000 was fulfilled with half that sum, as were all the others'. All statements of return on investment thereafter mislead; the real return always stood at twice the reported percentage. Flather's inflated evaluation aimed to justify this deception. Finally, the failed hope expressed for local participation indicated the extent to which Lowell industry remained the captive of outside interests, separated not only from the city itself, but remaining part of a financial/investment market without connection to the city's interests.[3]

Wellington, Sears pledged nearly half of the 1905 stock subscription. Far from having learned a lesson from the overdependence on the previous selling house, Dunbar not only promised it a 2 1/2 percent commission on sales, plus charges, in consideration for its participation, but also agreed that it could only be terminated through repurchase of its stock at par value. Revealingly, Dunbar wrote this agreement on the stationery of the selling house. Frederick Ayer, heavily involved in the patent medicine business in Lowell in earlier years and now significant in the American Woolen Company (organized by his son-in-law, William Wood), pledged $75,000 and was the second largest stockholder.

The *Lowell Daily Mail* announced that the mill would reopen at once: "It is understood that only a small proportion of the old operatives of the Boott have left town, so that when the gates are re-opened and the help summoned to take their old places, the mill will be able to resume operations with practically a full complement of operatives."[4] Ariel C. Thomas, the previous agent, would resume his post at Boott Mills (without the "Cotton" in the title).

REASSESSMENT

First as part of the decision to reorganize, and in a series of diagnostic efforts thereafter, the Boott brought a steady stream of consultants to analyze, describe, and prescribe for the ailing operation. Their commentary injected a note of caution, to say the least, into the initial discussions of the new owners. They drew these advisers from a substantial pool of respected authorities on textile and mill engineering, administration, personnel management, and related fields. From Boston, Providence, Lawrence, Fall River, New York, and the South they brought decades of tradition and significant experience and knowledge of factory operations. Hired by the directors, the consultants offered expertise, objectivity, and neutrality in case of an internal conflict. They reflected a situation in which mills were not only controlled by financiers without cotton production expertise but supervised by a new type of manager similarly ignorant of the factories' internal workings. The men in control lacked the ability to make crucial decisions without this assistance, a problem accentuated by a lack of shared opinions on goals and methods, as will be seen.

Lockwood-Greene and Company inspected the plant and operations during 1906–7. Its assessment initiated a litany which would continue throughout the mill's active life:

> We consider that when your Company took this property over it secured a manufacturing plant which was both as regards the style and construction of the buildings and the arrangement of the machinery and the relation of the processes to each other, a very poorly constructed and arranged plant, judged by modern standards. Mills 45' and 50' in width and six stories high cannot be considered good buildings for any department of a modern mill.

It recommended rebuilding elsewhere, or at minimum, that "all of the buildings with the exception of the storehouses and the No. 7 Mill should be destroyed and new buildings erected in their places."[5]

F. P. Sheldon of Providence, Rhode Island, commented similarly in 1907 and characterized operations and prospects:

> The power equipment of the mills is outrageous—the worst I have ever seen, as you doubtless know well. I think you must be losing fully 40 percent of all your power by friction losses on account of the arrangement of the driving. It would cost a very large amount of money, indeed, to reorganize this equipment, and I should hate to spend such a large amount on this equipment as well as on the machinery to adapt it to these old buildings. . . . There can be no doubt at all that the cost of producing goods in these mills must be very much higher than it is in good mills, and while it

> may not cause loss in years like the present one, still, when margins come down close as they have been in past years, it might be very hard to earn a profit which would warrant a large expenditure to be made on machinery and power.

The shafting system, like the buildings, had worn out. The types of problems he noted applied not only to the driving of the equipment but to nearly all aspects of the mill's operation. He cited unnecessary machinery poorly arranged, which required excessive numbers of workers to attend it. Excess floor space raised overhead costs and defied the close oversight "which is necessary for the best results." The buildings blocked thorough reorganization, so that changes in machinery and the division of labor, while helpful, would not place the Boott on a competitive footing with "a modern mill." Sheldon's diatribe became increasingly specific and focused, moving from power to machinery, to its layout, and culminating in oversight and labor. He advised creating a much smaller operation in a fraction of the space. These reports immediately confronted ownership with choices regarding the basic nature of their operation: they could live hand-to-mouth, with a goal of short-term success, or make the recommended changes with an eye to continuing manufacture.[6]

Revised Production

The extent to which the Boott had developed into an order mill augmented its difficulties. In the immediate aftermath of the takeover, high profits came from production of sheetings and drillings, with production complicated by the variety of fabrics that consultants noted. Yet despite the deficiencies described, Wellington, Sears moved the operation into finer combed goods and lighter duck by 1906. These costlier goods offered more profit to the selling house and offered a response to steadily mounting competition from more efficient mills, including southern ones, for coarse-goods business. As a result, seventy-six styles of cloth were produced at one point in 1908.[7] Tension resulted as market forces exerted more pressure in one direction than did mill conditions in the other. By the early teens production was strikingly diverse, with twelve distinct types of fabrics, each made at various weights and widths, in runs of all sizes. The mills poured out bags, Bedford cord, scrim, poplin, repp, twill, moire, corduroy, moleskin, velveteen, shelter tent, and duck in a total of ninety-three varieties in one week! Fabrics came and went as the market in general and orders in particular changed. They ranged from heavy ducks to fine lawns, and "even veiling with silk filling of equivalent cotton count of 350s" (gossamer).[8] Even the ducks included a number of materials of varying weight, while woven bags, toweling, corduroys, and numerous other materials, eventually even including sewn products such as curtains, contributed to the output. Production of forty different warp yarns and fifty filling yarns, combed and uncombed, appeared typical, in contrast to Lowell's image.

By 1914 the agent recommended dropping fine goods in favor of coarse and medium counts (8s to 16s and 24s to 32s). Despite that, corduroy, toweling, moles, and velvets predominated, along with various ducks, including shoe duck and automobile tire duck, a new product. Gradually toweling and scrim, materials which could be identified with the mill as its product (whereas much of their other cloth went elsewhere for finishing and/or stitching), grew in importance, as did tire duck by the twenties. Yet in 1923 weavers in #6 alone made forty-five different styles of cloth.[9]

Administration

The Treasurer

Frederick A. Flather now had nominal control of the mill, its employees, and its operations. He was a tall, lean man, his manner and outlook rigid, even cold. He exuded confidence and could be considered aloof. Yet his work did not come easily to him. Dr. Ghering, of Bethel, Maine, a widely known specialist in neurological problems, indicated he found Flather "constipated, nervous, and overworked" and recommended two months' vacation per year.[10] Similar complaints continued through his career and suggest a high level of internal struggle to maintain his position. He exhibited conflicting impulses, generosity toward distant family, but routinely tempered by moralizing, for example. Bequests of $100 apiece for long-time household employees carried the caveat, "unless they quit," demanding loyalty until his end despite years of service. Generosity and kindness, but judgmental and cautious—he was tension personified.[11]

He followed a path to the treasurership typical of Lowell: his father-in-law, Jacob Rogers, held a significant part of the stock of the new company, as he had of its predecessor (Flather married Rogers' daughter, Alice, in 1898).[12] Rogers had earlier characterized his son-in-law when an agent's position appeared: "He is about 35 years of age, in perfect health, perfect habits, absolute integrity and one of the most forceful, tireless, industrious men I have ever known. The moral hazard among mill men nowadays is confessedly great for there are many opportunities to receive commissions or favors not easily detected."[13] After more than seventy-five years of industry in Lowell, a director still displayed a primary concern for the company's ability to extract allegiance from an employee and valued it over technical competence even at the level of agent. In 1905, Rogers championed Flather for the treasurer's position in the reorganized Boott Mill. A series of telegrams, all to *Mrs.* Flather, in Chicago, described ongoing negotiations for the job amid great secrecy. They never mentioned Flather by name, nor was the job under discussion identified. The upshot was an offer at a salary of $7,500 a year. The secrecy aimed, as Mary Dunbar wrote her brother-in-law after the successful campaign, to avoid everything which could make the appointment and the salary "too Rogersy."[14] Nonetheless, Flather could apparently write without irony, fifteen years later, that he could not help another man get a job, "as my experience has been in a field where employe and employer were the only factors in obtaining employment."[15] His rigid decorum permitted him to ignore his own exception to his rule, and again indicated his habit of creating (and enduring) tension between warring interpretations of reality: between financial manipulation and faith in his scrupulous honesty; between generosity, family loyalty, and disapproval of his relatives' conduct; between his faith in advancement on his merits and his acceptance of nepotistic preference.

His path to this opportunity revealed much about the man and his new employers. Flather had initially worked at Flather and Company, the machine-tool–making firm founded by his father, in Nashua, New Hampshire. In 1887 he went to Pettee Machine Works, a major builder of textile machines in Newton Upper Falls, Massachusetts. He brought "not only some practical knowledge of machine designing, but a keen feeling for change which was not always found in the experienced and seasoned machinist" (which he was not). He rose to assistant superintendent and later superintendent of the works. His accomplishments there included several patented improvements, mostly to carding machines, but he took special pride in his labor management. He claimed to have averted a strike when he refused to accept a demand for sixty hours' pay for a fifty-eight-hour week. He threatened to close the plant during the installation of new machinery, then held a vote in which a continuation of the old pay system, presumably without layoffs during the equipment revision, won, 72 to 4. His approach interestingly combined a hard line with an impression of largesse: keeping the plant open and men employed. However, maintaining steady employment of the skilled workers at a machine-shop represented a long-standing practice designed to retain essential help.[16]

Blocked from rapid advancement beyond the position of superintendent at Pettee, Flather first returned to the family firm, then in 1893 became assistant to Charles Hildreth, the manager of the LMS. Apparently an impressive manager, within two years he was involved in unsuccessful negotiations with the McCormick Harvester company for a three-year contract starting at $3,500 per year.[17] At Lowell he helped with design and production problems, particularly with the comparatively new revolving-flat card. He made progress there not only with the immediate difficulties but also with the overall operation and equipment of the plant, bringing in new machine tools "by the freight car load." His contributions related to both machine design and mass-production techniques.[18] In a self-promotional letter to George F. Steele, manager of the Deering Harvester Company, Flather claimed to have reduced labor costs per card from $300 to $155 by early 1895, and later to $125. Well-traveled in these years while observing the foundry and machine-shop practices of others, in another letter to Steele he referred to his interest in the piecework practices being suggested for such work by his "friend Mr. [Frederick Winslow] Taylor." His interest in a system to give management control over skilled workers was in keeping with his lifelong attitude toward labor. In 1897, while Flather was on a trip for the LMS, Steele offered him $5,000 a year to replace him, superintending 3,900 men (according to his diary). Flather did not go to Chicago at this time.[19] Looking back, Flather cited the "first successful manufacture of worsted machinery in the United States" as his major accomplishment while at the LMS, but at the time he cited efforts at restructuring work and reducing labor costs.[20]

Flather ultimately did relocate to Chicago, working first as superintendent of the McCormick Reaper Works in 1901, then in a similar position for the International Harvester Company when it was formed through the merger of McCormick with Deering and others, finally as manager of the works there. Flather served during a stormy reorganization marked by struggles over control of sections of the new company and by bitter strikes in various branches over wages, hours, and the organization of work (skills and prerogatives). His advancement seemed to reflect success, but a letter from his father-in-law to his daughter Alice at the time of Flather's move to the Boott inserted a bit of mystery and doubt: he urged secrecy regarding Flather's election to the treasurer's job, "so people cannot for a moment suspect that he was out of a position with the Harvester Co. (which until the 15th of March he will not be) when he was elected."[21] Despite claims of success and complaints that he had not been to blame for failings, he had been fired. His impressive rise through the ranks of top machine-shop management had abruptly been derailed, with no employment in sight until the Boott opening appeared.

His early career was marked by intelligence, innovation, and ambition. He brought to the Boott not knowledge of cotton textile manufacturing but experience in managing men, planning operations, and introducing innovation. He had some knowledge of textile machinery production and had dealt with cotton manufacturers while promoting the machines LMS produced. A letter from agent A. C. Thomas to Flather some months after the mill had been taken over indicated the nature of their relationship and the location of manufacturing knowledge. Thomas defined the terms generally used in accounting the mill's weaving operation and described the manner in which the production per day and price per yard were calculated and which elements they included, such as oversight and labor.[22] Thomas conscientiously explained these simplest of calculations, the most rudimentary accounting, underlining the extent to which a treasurer could operate while divorced from an on-site analysis or understanding of the mill's operation.

Obviously, Flather was hired for executive ability and family connection, not textile knowledge. The agent, superintendents, overseers, and second and third hands provided that kind of supervision of the plant's operations, and the employees' skills produced cloth. But neither was Flather a treasurer of the old style, managing the mill from Boston with only occasional contact with activities in Lowell. On the contrary, he lived in Lowell and visited the mill nearly every morning, arriving early and then going to Boston on a mid-morning train to conduct cotton buying and relations with other mills, directors, bankers, and such.

Flather came to the Boott with firmly developed attitudes and a basic philosophy of financial conservatism. As early as 1892 he formulated a personal investment policy that ruled out purchasing stocks on margin, "gambling" on the market, or "buying anything which, if it proves worthless, will financially embarrass you." Furthermore, he continued, "I have made more money by honest labor than by speculation and in less time."[23] He applied a similar attitude to the Boott's affairs. His rules were, first, to accumulate and keep a large net quick (or liquid assets); second, to strengthen the mill's manufacturing position, in the North or South, but without endangering the net quick; and third, to broaden the market for the mill's goods. He was, according to *Textile World,* "ultra-conservative" in his approach to financial affairs.[24] Such a philosophy fit in well with that of his employers, and largely through their efforts he became a director of local railroads, banks, insurance companies, the Proprietors of Locks and Canals, and several other textile mills, including the Tremont and Suffolk, the Merrimack, and the Continental Mills of Lewiston, Maine.

His occasional diaries demonstrated the application of his philosophy to daily operations. In 1910 he noted he had told the agent "that the tearing out of the old engine room did not look safe and that I had plenty of money to spend for safety."[25] Another time he "sent word . . . that idle men in 6-2 tower [mill #6, floor #2] cost money and is result of poorly instructed and placed supervisor." Nor were his help in the Counting House free from problems: "Holgate was anxious about his health if he remains in his room, which vibrates and is noisy. Told him we all had to put up with something and I was not interested in that question (his real reason I think is that he wants Thomas' office)" [the agent was away, ill].[26] Flather's peremptory and suspicious response to Holgate revealed another aspect of his character, a lack of felicity in human relations.

The picture that emerges from Flather's diaries reveals a man of meticulous manner and devotion to his task, able to play a role of power on the one hand (when at the mill) and subservience on the other (when reporting to the directors). He carried an air of competence, but not tact. This view is reinforced by a note of his upon receiving the news that Boston's elite Union Club had determined not to grant him membership on account of a reputation of "not getting on well with people." Evidence from as far back as his time at the Pettee Works apparently played a role. "To me," he wrote, "it is praise. I do not use my *position for self* [his emphasis]." He went on, "[I am] proud of the things that I have suffered in doing for my employers."[27] With his self-righteousness, combined with his ability and ambition, he had not made friends of all those he had met on his way up. The directors had installed a man of dignity and executive reputation who would represent their interests faithfully. His view of his position combined with his temperament to leave him personally isolated but thoroughly dedicated to his mission.

The Agents

Investors might discern opportunity for profit, and the treasurer might manage the finances, but if the agent could not coordinate, direct, and evaluate production and communicate his findings effectively, the others could not hope to succeed. During the early Flather years, Ariel C. Thomas (1905–8), John Whitten (1908–10), E. W. Thomas (1910–20), Fred Lacey (1920–21), and Benjamin Holgate (1921–?) served in the position of agent. The brevity of the terms of these men, several of whom died in office, suggests that this stressful position carried its own occupational hazards.

Flather described the first agent brought in by the new company: "Wellington, Sears having a man, John Whitten, under contract but on the loose pulley, made him Agent of Boott. He divided his time between Manchester, where he lived, a desk in Wellington-Sears office, Suncook Mills and Boott Mills. He broke down and at end of his second year died."[28] Following him came E. W. Thomas, formerly agent of the Tremont and Suffolk and therefore well known to the Ayers, who controlled that operation.[29]

After Thomas died in 1920, Flather recommended the appointment of Fred Lacey. His description of the Bradford, England-born Lacey as coming from "mill people" led into an account of his training in textiles in the mills of Massachusetts and Montreal as he rose to the level of superintendent at the Boott.[30] Flather's evaluation said much about his attitude, the state of American textile knowledge, and factors influencing the Boott's direction. Although the cotton textile industry had operated on a large scale in this country since the 1820s and reached maturity long before 1905, England still stood as the land of experts, the place where agents were born. While Lacey's training occurred in the United States and Canada, his lineage still deserved mention. His cotton education had followed the typical nineteenth-century path, in the mills, and his jobs were many and varied. By moving from mill to mill, he matched his talent to his ambition as he rose to slots of increasing authority, and he learned from comparisons between each of the mills at which he worked. Although mobile, both geographically and upwardly, he maintained a reputation for reliability and neutrality: he knew his place and did not interfere or take sides in the disputes of his superiors. His understanding of the cotton process emerged from experience at performing the tasks associated with it. His allegiance was to the mill and its administrators, his knowledge tied to the work experience.

Like Whitten and Thomas, Lacey did not long survive the job; he died the next year, July 30, 1921. Flather's recommendation of Holgate as Lacey's successor indicated the extent to which the management thinking had changed during his administration. Holgate's career provided an indication of a new direction:

> In 1910 he took up the maintenance of manufacturing schedules, made the promises of deliveries to selling house and made the estimates upon which selling prices were based. He has designed, or analyzed, and organized the fabrics that we have made since 1910, has handled the complaints from customer, selling house, or from our departments, and stuck to them until satisfactorily disposed of.[31]

The new man was experienced in office work, only slightly in shop-floor activity. He learned fabric analysis while working in the office, studied the cotton process in school, and proved himself in areas connected to design, sales, and customer relations. He demonstrated abilities to handle the "customers' peculiarities" and captured the valuable Woolworth toweling account after the selling house failed to do so.

The new agent was fitted to a growing trend in management to rely on careful analysis of figures to plan and evaluate a mill's program, work assignments, and sales plans. Mill-floor learning and knowledge had been displaced. Holgate's installation reflected Flather's allegiance to the new school.[32]

These men, the agents, regardless of their backgrounds, still ran the mill. The series of deaths in these years raises serious questions about the level of intensity in an operation attempting to avoid the fate of its immediate predecessor without recommended overhauls.

CONFLICT WITHIN MANAGEMENT

At several points in Flather's accounts of the Boott's activities (above), intrusions by Wellington, Sears surfaced. Their interventions must have been difficult for him to protest, impossible for him to resist. On the other hand, he had presented them with unusual opportunities to interfere and to assume the worst about his knowledge. For example, early in 1906 he wrote to them inquiring about the proper machinery to run for planned production. Their reply recommended equipment from spinning through weaving in excruciating detail. His request revealed a lack of expertise, despite the presence of an experienced agent; their reply demonstrated a complete lack of confidence in him.[33]

Wellington, Sears' distrust of Flather grew. Just after Christmas, 1907, he noted in his diary that the selling house "proposes [Agent] Whitten to take charge of mill absolutely and I to have no desk and no business or authority at mill." By January first it was "done," and Flather had been humiliatingly displaced.[34] Within a month Whitten recommended tearing down the mill, which suggested that whatever difficulties led to his appointment did not disappear after his arrival.[35] Whitten died after his second year as agent without visible impact on the mill's condition or production.

The disagreement over power and direction which had simmered since 1905 became unavoidable in 1914 when the directors considered plans to increase investment in the mill. In pondering how to improve the Boott's operation and competitive position, some board members favored raising $600,000 through a stock issue to underwrite past and future improvements to both machinery and plant.[36] Wellington, Sears, on the other hand, touched off a bitter and often personal dispute when they charged that Flather should be ousted on account of his "extravagant" management. They claimed that he was to blame for the fact that "the goods were costing so high that they were almost prohibitive to go into the market to sell and make a profit for the mill."[37] They found the profit of $774,000 over the past eight years, a return of about 35 percent annually on the $300,000 investment, insufficient and the result of poor management.[38] The mill could not continue until the responsibility for its "difficulties" could be placed.

Jockeying for position took several forms. In January 1914, the directors created a committee to investigate the business. In March the selling house resigned, a move requiring it to be repaid its full investment. In May the board refused it permission to rescind its resignation.[39]

The composition of the board of directors had changed somewhat since 1905. Wellington, Sears still held the largest single block of shares, 2,300, but Albert F. Bemis, owner of several other substantial cotton mills, had become second with 1,992, and Charles F. Ayer third with 1,656 (Flather owned 250). Bemis and Frederick Ayer, who together exceeded Wellington, Sears, in power, had been supporters of Flather, but Bemis had played his role by proxy while spending most of his time in China. When he returned, in 1914, the possibility arose of his shifting his support to the selling house faction.[40] Although Flather later denied going to California in order to meet Bemis's ship before Wellington, Sears could talk to him, he also acknowledged the story was "approximately true."[41]

To settle the dispute over the mill's management, the directors appointed Cumnock, now treasurer of the Appleton Company, to examine the running of the mill. (An effort was made to speed the inquiry in order that the conflict could be settled before Frederick Ayer sailed for the Mediterranean on his yacht.) Flather's responses to a letter from Bemis in June gave his version of the mill's direction:

> Before January 1, 1907, all orders and inquiries were passed upon by me. Beginning with January 1, 1907, and during his incumbency, orders were passed upon exclusively by Mr. Whitten, if at all by any direct representative of the Mill. This practice developed as the result of a policy which Messrs. Wellington, Sears & Company wished enforced. Since January 1, 1910, and until a comparatively recent date the same policy has been continued by the Selling Agents with Mr. Thomas with rare and spasmodic exceptions. . . . I have not approved of the character of the business, (with exceptions), furnished the Boott Mills, and have as often as seemed practicable made this fact known to Wellington, Sears & Company. By my direct and amicable dealings with them I hoped we could mutually solve our selling and manufacturing problems.

From the treasurer's point of view, direction of the mill had been taken from him by the selling house, which ran it without regard to types, capacities, and conditions of machines. The primary stockholder dictated production. As a final economy measure, it had stripped the plant of every expendable employee in a massive layoff.[42] Flather could only counterpose amicability to their power.

Cumnock prepared a report that supported Flather and was quoted at the pivotal director's meeting:

> These policies have been determined by the selling house and not by the mill treasurer, and if it is true and the selling house has been directing the manufacture of certain goods and a great many varieties of a great many kinds of goods and sizes and threads, with the result that now there is on hand some $450,000 worth of those goods manufactured according to the directions of the selling house which are not sold and apparently cannot be sold immediately, why you may put two and two together and find out where the shoe has pinched.

The unsold inventory stood as a condemnation not only of the advice given the Boott, but also of the temerity which had led Wellington, Sears to take over its direction. Flather's victory was not complete, however, nor his endorsement thorough. He was given two years to shape up the operation he had been hired to run nearly ten years before.[43]

Cumnock found the mill "well maintained," an evaluation apparently made within the parameters of its traditional policies. In other words, maintained the way it had been under his supervision at the time R. G. Dun criticized it. He also noted the addition of some new machinery over the previous ten years. But he felt production at the Boott should be narrowed to an average of 24s yarn, and

> that it should promptly greatly reduce its present variety of upwards of thirty different warp numbers ranging from 7s to 68s, and twenty-six different numbers of filling yarns ranging from 9s to 55s. Such a large number and wide range of yarns as you are now making means a high manufacturing cost and must make it hard to get and hold an established business. It is also one of the principal causes which lowers the productive capacity of the Mill which during the past year I have proved was only 64.72 percent.[44]

Cumnock's warnings about the desirability of producing long runs of coarser yarns to increase efficiency, while seeking wide distribution under the company's name, proposed a combination of productivity through management planning and success through marketing. His testimony was instrumental in saving Flather (and indirectly initiating the new investment program). Significantly, his complaints echoed those of Parker in 1902 and Sheldon in 1907 (and presaged those of Robert Valentine in 1916 and the Textile Development Corporation [TDC] in 1928).

Managing Production

The complexity of managing several interrelated production steps taking place continuously in the many rooms of the several buildings and the interplay of people, machines, cotton, and markets required precise measurement, analysis, and direction. Workers in the Counting House provided the basis for the coordination necessary for this system.

Successive consultants provided analyses of office procedure as a key ingredient for the Boott's future success. Their advice was the more welcome in the light of other comments indicating that nothing could save the plant itself. If new practices, rather than new machines or new buildings, were the key to success, it might be accomplished without great expenditure. When L. A. Hackett reviewed operations in 1911, he criticized the treasurer's practice of keeping his office in Boston rather than at the mill ("modern mill practice") and of using a selling agent rather than selling direct. Wellington, Sears' role made change impossible, of course.

Hackett described the administrative structure and found it wanting. The agent dealt with every department of the mill and oversaw a superintendent of the works, the master mechanic, and the office manager for the three main divisions of the operation. The superintendent of the works directed production through departments (Cloth, Picking, Carding, Ring Spinning, Mule Spinning, Spooling, Twisting, Dressing, and Weaving.) The master mechanic managed "Yard Men, Machine Shop, Power House, Carpenter Shop, and Supply Room." The office manager superintended the "Paymaster, Cost Man, Bookkeepers, Messenger, and Stenographers." The diffusion of authority led to a lack of coordination and of accountability. Information flowing from the mills lacked utility. Accounting for power consumed, supplies used, or other aspects of the "burden expense" (the costs associated with production in a given room) appeared speculative at best. The office could not make the assignments, analysis, and predictions which were needed, nor could the overseers measure the degree of success in their own rooms.[45]

In the aftermath of the 1914 battle for control of the Boott, Flather made efforts to improve management's ability to monitor and improve production control. He instituted a new cost accounting system, and in 1916 he "told Reed and Holgate to place all they were doing on other subordinates and try to do something in their own way to get the Boott out of its orbit."[46] Such efforts offered limited impact and did not enable this management to influence the course of cotton production sufficiently. The need for expertise in mill management brought further advisers.

In 1916 the New York firm of Valentine, Tead and Gregg, Industrial Counselors, studied the Boott's operation. The resulting Valentine Report was noteworthy for its attention to the role of the human element in the operation of the mill. It reflected a humanitarian approach to labor management, and asserted that heightened productivity could result from moral and practical improvement in treatment of the workers and increased attention to their relationships to management and the factory. Robert G. Valentine, head of the firm, played a leading role in this school of industrial counseling. He served on the Massachusetts Committee on Unemployment, a group seeking to approach problems of labor unrest and poverty by addressing the difficulties associated with the intermittent nature of work.[47] His firm took a different and broader view of techniques for increasing productivity than did Flather's old friend Taylor and his followers.

The Valentine Report faulted the decentralization of power at the level of both treasurer and agent for creating "confusion, constant shifting of policy, and delay." It suggested that despite the five years since Hackett's visit, oversight remained in disarray. Sales and Planning Departments determined products and scheduled their manufacture, respectively. The Boott's market orientation had created a situation in which Sales directed an operation it did not understand.

Scheduling represented the sole opportunity for creating and maintaining the necessary balance between the production of the many departments to supply the needed materials in the proper quantities for uninterrupted work. The shifting needs of an order mill heightened the need for precise direction and control: "Scheduling is rendered very difficult by the frequent changes in orders received from the sales department. Overseers are occasionally asked by the planning staff to deliver quantities that are beyond the capacity of the equipment at their disposal. This is unfortunate and causes some resentment."[48] As late as 1927 a memo noted that the office "has never attempted to schedule the filling but is working on plan for doing so."[49] That Sales retained its power without an increased understanding of the mills' capabilities indicated little had been gained from the ouster of Wellington, Sears.

The Valentine Report also faulted the administration for lacking personnel (the superintendent worked without benefit of a clerk to "handle routine papers, keep records, and answer the telephone") and misplacing help (clerks for the overseers of the various departments were found in the Counting House: "Overseers wanting to give them work have to leave their rooms and go to the office"). Overseers showed little interest in the accountant's cost figures.[50] Their difficulty in dealing with the figures involved becomes easier to understand, the accountant's work easier to appreciate, and the process for figuring costs more meaningful if we look at weaving costs as an example. The accountant took hours from the loom hours book and production from the cloth room account, and then calculated related payroll amounts. Indirect labor included the "supervisor, inspector, burlers, fixers, oilers, bobbin boys, filling carriers, truckers, room girls, battery hands, scrubbers, sweepers, cleaners, supply man, elevator men, and misc." Direct labor included loom changers assigned to the style of cloth being woven. Pounds (of cloth made) or picks (of the shuttle) were added to the payroll and calculated according to the style. On styles for which piecework rates were based on picks, pounds were estimated by dividing picks per yard to get total yards, then dividing by weight to get pounds. These results were then posted to labor distribution sheets at the end of the month.[51] This procedure for determining the costs of weaving describes only one aspect, labor, of one operation, weaving, and had to be repeated for every process, every material, and every style to check its cost for the mill. The complexity and occasional nature of this accounting process made careful calculation of expenses for the many styles of cloth difficult, if not impossible. Despite operating as an order mill, the Boott behaved in many ways in the manner of its earlier configuration as a mass-producer of staples.[52]

For basic accounting, the office moved into the mechanical age by 1913, when it purchased tabulating machines such as the "Millionaire."[53] For some time thereafter new office technology seemed to lag behind earlier practice: "Calculating machines break down and delay of getting them repaired at once suggests impracticability of their use for us. Our girls actually went through pay-roll faster by hand than with change machines."[54] Old methods and expectations did not move gracefully aside in any aspect of the Counting House operation.

These problems, and their continuance, meant more than inconvenience: "All this results in friction, lost time, lost energy, dulled initiative, indifference, impatience and discouragement. The tone of the main office force is lax. It is reflected in the tone of the whole mill. The workers suffer in morale, wages and poor working conditions."[55] Disorder and frustration flowed from the office through the mills.

In keeping with Flather's self-denying and frugal nature, however, the offices in which the new levels of activity took place did not change and appeared plain: "The office would greatly benefit by fresh paint and calcimine. At present it is unnecessarily cheerless. To callers at the mill it gives an appearance of indifference to appearance and lack of prosperity that is most unfortunate."[56] Flather's ascetic approach to his surroundings fit his sense that the success or failure represented a test of will, as he revealed in an interview:

Dillaway: "Why do some mills pay, and why do some others fail?"
Flather: "Well, why do some people fall down, and do others stand up?"

His attitude indicated a belief that success followed individual effort, that the mill's course reflected his ability to impose his drive, his character, on the operation. Such a belief required Flather to separate rational evaluation of the operation (as reflected in the reports of outside consultants) from an identification with the Boott which made its success or failure a part and result of his own character. This personalization presented an enormous moral burden.[57]

The final, and most significant, change in the management at this time was the addition of two new workers who would play an increasingly important role in the mill: Flather's sons, John Rogers and Frederick. The arrival of the brothers signalled the initiation of a voluminous record of mill operations, a rare diary of daily events which contributes greatly to our ability to discover the story of the operation of which the young Flathers had become a part. In this way they made a significant, unique contribution to modern textile history. Born a year apart on the same date (in 1899 and 1900, respectively) and raised as twins, the two operated as a team within the Boott and served their father's (and family's) cause for over thirty years. They maintained an involvement in all aspects of the Boott's operation but gradually concentrated their efforts to a degree: John Rogers on marketing, sales, and personnel, Frederick on technology and operations.[58] Both had worked in the mill at times previously and came to the mill simultaneously after graduating from Harvard. Frederick majored in engineering while John Rogers, the elder, had utilized his extra time to attend Harvard Business School. Frederick continued his studies at Lowell Textile Institute evenings from 1923 to 1925, adding textile engineering to his Harvard foundation.

Their nearly immediate installation in directorships of organizations such as the First Congregational Church, the Lowell Institute for Savings (where Frederick represented a fourth generation of the family to serve), the Chamber of Commerce, and, for John Rogers, as Scoutmaster for the Boy Scout troop reveals their status in the Lowell community.

They occupied special but less clearly defined positions in the mill. John Rogers noted, for example, that "I am receiving no pay for a few months so that I may be more free to ask questions and look about." He studied the various departments and noted the course of his education in a manner which revealed the novelty of the topics to him. He hoped to be able "eventually [to] establish new and money-saving plans." He also desired to divert attention from any Boott mistakes, a tactic he associated with his father. Presumably the directors were enough of a problem without having undetected errors revealed to them. He identified and aimed to mimic his father's devotion to duty and willingness to perform the tasks set by his supervisors as the path to success.[59]

Chapter Seven

Technology and Labor

Responding to Change

Production was reverting to the Fine Mill, making the lighter weight yarns and fabrics, including corduroy, and the Coarse Mill, producing the heavy ducks and bags made in part from the waste from the first operation. Furthermore, individual buildings were being converted to narrower functions; for example, #6 made only fine goods, but was also restricted to spinning and weaving only. Reorganization promised a more coherent pattern of operation without new construction.

The Boott's revised production required a new Testing Department. Yarn and cloth from the various mills came here for assessment of tensile strength, consistency, and conformity to standards. The lab compared swatches of the finished product to the customer's sample that the mill had attempted to duplicate. In one case, 60,000 yards produced to match a small sample had resulted in the return by the customer of 30,000 yards as unacceptable, a situation the mill needed to avoid. Strength testers, twist counters, and related devices enabled management to plan production and to measure the extent to which it succeeded.[1] This meticulous, at times microscopic, work stood in stark contrast to the mill's general mechanical operations, but at the same time it demonstrated the preciseness of result for which the mill needed to strive to succeed as an order mill. In other words, while the textile machines appear primitive according to modern computerized standards (and eyes accustomed to such), they had to produce results at a level of precision that required minute analysis to determine the extent to which a standard had been achieved or missed. This great hulking machine of a mill created precisely defined and measured products.

The machinery in place at the turn of the century did not facilitate this process. The fixed-flat cards installed in the 1870s, along with the railway heads accompanying them, were on the verge of obsolescence when they were installed and had been thoroughly out of date for decades by 1905. Recognizing this situation, management finally began purchasing the next generation of equipment, revolving-flat cards. These increased production and saved on labor. The old cards sold as junk.

While nothing on the order of the earlier introduction of automatic looms occurred, the Boott quickly recognized the advantages of certain new inventions. During 1906 they bought Barber hand-held knot-tying machines. They were reputed to tie better knots, to save in spooling and warping, and to produce fewer imperfections in the cloth: "Last but not least, the knotter not only reduces the labor cost by saving time, but also makes possible the employment of less skilled labor by doing the work which required most skill on the part of the operative."[2]

While seemingly minor in themselves, these hand-held devices followed the usual trend, and goal, of textile innovations by utilizing less of a person's capability and thus permitting the hiring of less expensive, more easily replaceable, labor. They were also typical in their pyramiding effects, as improvement in the accomplishment of one task led to better quality work at succeeding stages of production, thus permitting the speeding up of work there, as well. The drawback to this "progress" lay in the fact that work became less skilled, more monotonous, yet more stressful, because of the new pace.

The Boott also quickly adopted the new Barber warp-tying machine, first available in 1904. This machine connected the tail of one warp, still in the loom's harnesses, to the beginning of a new warp of identical pattern: "It ties about two hundred and fifty knots per minute, and does the work of *twenty girls*. Drawing in by hand had always been a relatively heavy expense to the manufacturer: by the use of this machine the labor cost is cut down two-thirds"[3] [emphasis added]. This machine also speeded production while eliminating skill and jobs.

The drawing-in machine represented the third of the new Barber developments brought to the Boott early in the reign of the new administration. This machine pulled the individual warp threads through the eyes of the heddles (which were manipulated by the harnesses to form the pattern or construction of the cloth) and served the usual ends: "Yet it does effect a marked saving over hand labor, since one man operating a drawing-in machine will draw in about six times as many warps per day as a girl can draw in by hand on the same grade of goods." Melvin T. Copeland described the work eliminated by the drawing-in machine as "a severe strain upon the mill workers."[4] However, drawing-in had long been known as a job to which the operatives tried to move on account of the clean, quiet atmosphere of the job in comparison to others in the mill.[5] Only extreme increases in pressure for speed could have changed it into a "severe strain."

Attempting to make the most of the equipment and space available without significant expenditure meant a constant juggling of machinery from one place to another, purchase of new technology in bits and pieces (which often perpetuated the problems it was intended to solve), and installation of various attachments to machines to bring advantageous new developments to the mill without acquiring expensive new equipment. The treasurer approached the directors with proposals for improvements, but either from innate caution, limited expectations, or belief that the plant itself did not merit thorough re-equipping, he moved hesitantly. During a time of little innovation in textile technology such a plan could maintain competitiveness. Flather's modest proposal in 1923 revealed his approach. He wanted to replace about 100 of the 400 LMS looms:

> 112 Draper looms would require 11 weavers, in place of 22 weavers at present. The looms produce about 22,400 yards per week. The cost of weaving could be reduced one cent per yard, or $11,000 per year. The Treasurer does not advise the purchase of 700 looms to balance the mill, which might cost $250,000. For the present authorization is desired to purchase not more than 200 looms, either new or second-hand involving not over $70,000, if and when market seems to demand them.[6]

Flather listed seriously outmoded looms, both too narrow and nonautomatic, stressed the limited nature of his proposal and its advantages, and contrasted his proposal with the far more costly alternative recommended by consultants. Although his mill was lacking looms of needed width and employing twice as many weavers as necessary, he hesitated before his directors. Purchases brought some new machinery for the production of fine goods—Model K Draper looms, for example, which would manipulate twenty-four harnesses, and new warp-beam heads for velvet production—a series of comparatively minor purchases. The mill's capacity to produce finer and more varied materials was being steadily, if slightly, augmented despite the advice of various consultants to stick to a few coarse products.[7]

These changes represented small investments and reflected immediate needs, as when the Boott bought thirty-six twisters to fill a large government order on one occasion, acquired seventy-two secondhand looms for another, and widened looms for a third job. In fact, allowances for machinery replacement in the years after the Boott Mill's reorganization ran as low as $70,000 at the start, when little equipment had been replaced in over twenty years (even the minor expense of bobbin-strippers was postponed until the advent of three-shift operation in 1927). Most purchases were concentrated in the early teens as equipment from the nineteenth century wore out. Even after that time, well over a third of the machinery still predated the Boott Mill takeover. During 1920 and 1921, more money ($100,000) was spent on advertising and merchandising than on equipment.[8] While the directors generally opposed reinvestment, Bemis demonstrated the exception by endorsing a recommendation in 1912 to spend $250,000, primarily on turbines, and urging the purchase of vacuum strippers for cards in 1917. The justification of the latter exhibited the company's thinking: "This is to eliminate dust, increase product 5 percent and decrease cost 25 percent. At 15 percent for interest and depreciation, the further saving is estimated to amount to $787.50 per year or $3,515.40."[9] Benefits for investors had to be immediate and substantial; environmental benefits for employees received no mention. In keeping with this philosophy, they easily justified replacements of worn-out water turbines by citing the potential savings in power costs to be realized, on the order of $30,000 per year for daytime operation alone in one case.[10]

A list of parts purchased during six months in 1913 suggested some of the activity related to equipment which was ongoing. Looms, for example, required temples, shuttles, picker sticks, binders, and crank shafts. New pulleys, hangers, bushings, and bearings helped maintain the power transmission system Loper had condemned. The spinning rooms needed new separators (to keep the yarn being spun on one spindle from becoming entangled with that on the next), cutter blades, chain, top roll springs, bolsters, nails and bolts, brackets, studs, spindles, and filling cutters. They installed new piping and valves, repaired the adding machine and the two-horse truck, purchased new parts for the steamers and elevators, bought a new die, fixed sprinklers, and at the same time acquired unlisted items, of less than $2.00 value each, which amounted to $7,696.83. The situation was purposefully made the more complicated by the machinery companies: spinning frames, for example, installed as late as 1910 lacked "perfect interchangeability of parts," and so repairs were an individual matter and required replacement parts from the maker, for whom they represented a major source of profit.[11]

Electrification of the drive system began, hesitantly, in 1912. Motors did not power individual machines but turned the existing shafting; pulleys and belts continued to transmit the power. The combination of foresight and hesitancy, insight hampered by partial implementation, appeared again and again. In 1927 the Boott experimented with long-draft spinning, which indicated an ability to recognize the potential of such an advance, but at the same time called attention to an inability to come to terms with it or to get the plant working properly with regard to known, long-term difficulties. Long-draft spinning represented one of the three major advances in cotton technology in a period of 100 years (the high-speed spindle and the automatic loom being the other two):

> Long-draft spinning and roving was nothing more nor less than realization of an ancient ambition to reduce the number of steps through which yarn was processed to reach its final size and strength. If the drafting capacity of the spinning frame could be increased, one or more of the roving frames employed in the processing which preceded spinning might be eliminated.[12]

Problems associated with spinning made the attempts to produce greater drafting very difficult to solve. The fibers passing through the drafting rolls had to be under control in order to maintain a given weight of product, but they could not be drafted more rapidly unless they were sufficiently free from control to accept the greater attenuation, a seeming contradiction. Fernando Casablancas, a cotton mill superintendent working near Barcelona, Spain, devised a system in 1913 that would control the fibers while using short endless belts as part of the drafting system, permitting greatly increased drafting, and eliminating one of the three roving frames while producing higher quality yarn. Although World War I delayed the spread of this invention, its potential was immediately recognized and various machine builders worked to develop their own versions.

The two younger Flathers traveled to the Boston Manufacturing Company mill in Waltham in 1925 to view the Casablancas system in action. The maker then offered the Boott a low quote on several thousand spindles as part of an effort to initiate pilot operations of significant size in an effort to encourage adoption of the new machinery on a wide scale. Two years passed before even a token installation took place.[13]

Experimenting with such a novel device in 1927 in this country was still a progressive step and paralleled prompt attention to other major innovations. On the other hand, even this small experiment was hampered by existing conditions in the mill. In 1928 they made two attempts to install a new humidification regulation system to facilitate the operation of the long-draft trial. The inability to maintain the steady high humidity required for efficient spinning (and weaving) indicated the failure of previous such efforts. In fact, a report in 1928 showed that relative humidity in weave rooms ranged from 42 to 82 percent on a given day. Problems also persisted regarding the use of the Draper automatic looms. While the Boott had been quick to adopt them, in 1928 belts still drove them all and at speeds of just 142–52 picks per minute, speeds that were improperly inconsistent and well below acceptable levels.[14]

CONDITION OF MACHINERY

Both employees and consultants lodged complaints about the poor condition of the machinery which affected managers, fixers, and operators alike. The "Textile Development Corporation Report" (TDCR) in 1928 scrutinized every loom in terms of a forty-one-point checklist. It found, for example, that brakes were "not set properly," gears were dry, and bobbins were of the wrong size and caused waste.[15] The loom-fixers employed inconsistent practices, as did management, resulting in excess waste, loom stops, unnecessary difficulties, and reduced earnings for the weavers. This review of machinery and conditions also examined every spinning frame in the mill and critiqued them individually. It suggested the following: "Correcting of machine conditions, especially steel rolls, spindles, cap bar nebs, thread guides, builder cams, rings, front saddles, and the same driving pulleys for the same yarn number." It noted slippage of dirty belts and recommended cleaning and dressing, very simple procedures which consultants should not have to address.[16]

The machines' condition could also endanger those working with them. The Valentine Report drew attention to the hazards of both exposed gearing and belt drives in 1916, noting that the latter "adds to the workers' sense of insecurity and hence decreases their liking for the work," on account of the ease with which a worker or his or her clothing might be caught in the belts.[17] An employee noted in 1926 that one group of spinning frames was "open head and dangerous" because of their exposed gearing. Those frames were thrown out later in the year.[18]

Narrow aisles between machines and crowded conditions in the mills also affected the workers: they blocked the free flow of work, made cleaning and repairing difficult, and prevented proper lighting. Wider aisles would have improved trucking of materials, changing and repairing machines, doffing, and cleaning.[19] Such rearrangement posed difficulties without individual motors since it required major reorganization of the system of shafting and pulleys as well as the machines.

The main problem with the machinery was its age. Insurance appraisals of Boott equipment done for 1905, 1922, and 1930 show that they ran considerable quantities of outmoded, worn-out equipment. Top-flat cards, spinning frames, and looms from the 1870s persisted in the early part of the period, well-worn machines predominated by the twenties, and nothing was purchased between 1922 and 1930.[20]

An appraisal by the Saco-Lowell Shops in 1931 found the opening equipment modern, but its layout obsolete; picking, mostly obsolete; carding, all more or less obsolete (new machinery would increase production and quality while reducing cost); combing, also obsolete, while new would cut production costs in half; drawing machinery also obsolete, as was roving equipment, thirty-one years old, and spinning, without long-draft. Spooling and warping they found too antique to deserve comment. While they did not assess (or sell) weaving equipment, they observed that the improvements they could offer would "help greatly in the weave room."[21]

The adoption of certain new technologies and experimentation with others make it clear that the mill's managers knew of current trends but pursued them selectively. The refusal of the directors to make wholesale changes and the inability of the mill to fully utilize the benefits offered by changes they did introduce reveal an approach limited by the amounts of investment they were willing to make. Even where labor costs were highest, in weaving, they only partially adopted automatic looms and, since what they had was belt-driven at inconsistent speeds, they only partially utilized their advantages. They quickly purchased inexpensive knot-tiers but only experimented with long-draft spinning, and mill conditions blunted its effect. Similarly, electrification meant driving shafts (still turning old-style belts and pulleys) electrically for many years before they began to motorize individual machines. Thus the pursuit of automaticity continued in minor ways, but the production equipment indicated that the new owners did not intend to pursue profits by investing in new technology. By 1930, some departments still ran with virtually no equipment purchased by the new owners; old and new machinery was generally evenly divided in the rest.

The Role of Factory Buildings

When the new owners took over the Boott Mills in 1905, they inherited buildings, machines and their arrangement, as well as existing relationships among workers, managers, and their environment. The buildings had grown incrementally, work had developed in relation to changing machinery and traditions, and the overall organization had evolved not only from management choices, but also as an accretion of innumerable events, a product of the history of the mill and of the industry, including the long-time interaction of management and labor.

Nowhere was the legacy of the old corporation, the Boott Cotton Mills, more apparent than in the buildings. The plant included virtually the entire original fabric dating from 1836 and thereafter. No manufacturing building had ever been torn down, and none of consequence was built after 1880.

Pickers transformed baled cotton into laps, sheets of fiber wrapped around sticks. Fire danger led to location of this process in separate buildings. Courtesy the Flather Collection, Lowell Museum.

Revolving-flat carding machines further cleaned and parallelized fiber, through action of wiry card clothing, and produced sliver, coiled into cans. Men did most of the work in these first two categories. Courtesy the Philip Chaput Collection.

Drawing frames combined four slivers into one, made smaller through drafting, again coiled into cans. Women predominated in the process of drawing through ring spinning. Courtesy the Flather Collection, Lowell Museum.

Slubbing frames continued doubling and drafting, winding up the product with a bobbin and flyer mechanism. Courtesy the Flather Collection, Lowell Museum.

Jack frames added to doublings, which mounted to the thousands in order to increase the evenness of the product by averaging out inconsistencies. Courtesy the Flather Collection, Lowell Museum.

Ring-spinning frames completed drafting and doubling, twisting the fibers together into yarn, generally for warps in looms. Courtesy the Estate of Frederick Flather.

Complex spinning mules made loftier weft yarn. Skilled mulespinners were the only group of workers to appear in this 1911 series depicting the entire production process. Courtesy the Flather Collection, Lowell Museum.

Well-lit drawing-in stands held warps while women drew yarns through harnesses for weaving. This intense but quiet work was among the most desirable available to them. Courtesy the Museum of American Textile History.

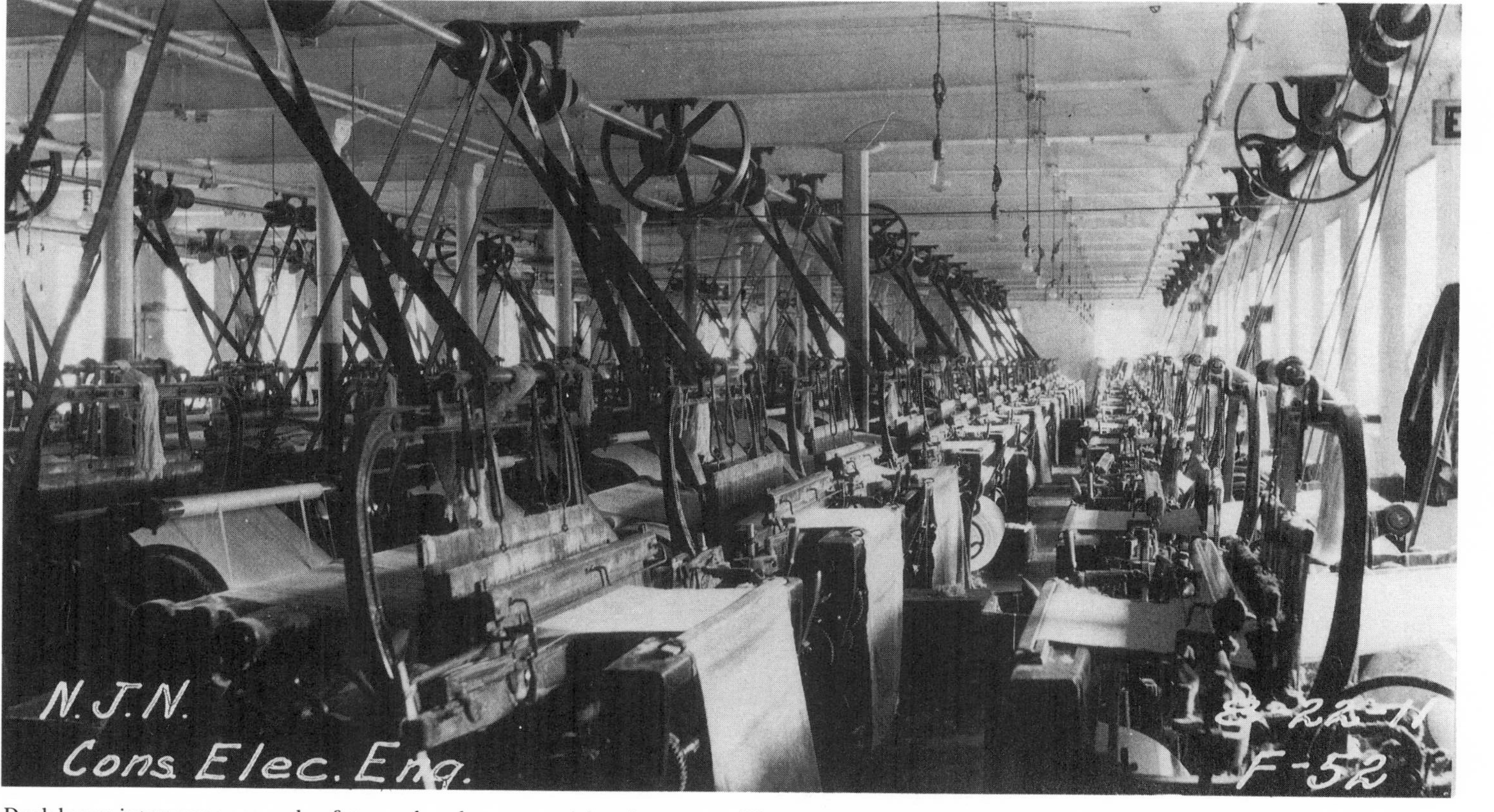

Duck looms interwove warp and weft to produce heavy material such as canvas. Weavers could tend only a few looms while keeping shuttles supplied with weft. Courtesy the Flather Collection, Lowell Museum.

Draper "automatic" looms with batteries of bobbins kept shuttles supplied and weavers could run about three times as many looms. Dobby motions on these "K" Models produced complex patterns. Courtesy the Philip Chaput Collection.

Management spent small sums on fixing floors and foundation piers, out of necessity, but despite the treasurer's awareness of "weakness," for example in the #5 mill, funds for repairs were not forthcoming. In 1917, Bemis asked for a professional examination of the buildings before committing himself to improvements. Expenditures came in areas producing immediate savings, as in power. They could find $80,000 in 1927 to "enlarge the coal pocket and save on rehandling." On the other hand, throughout this period the looms shook the structures in a manner not only annoying to the operatives but also threatening to the integrity of the buildings. Charles T. Main Company recommended reorienting the looms in mill #6 to protect the building.[22]

In 1920 Flather displayed a painful awareness of the situation: "We have an intolerable situation caused by four improper existing walls, the lack of one new wall, the lack of an elevator, poor lighting, poor toilets, poor safety, poor communication, poor transportation. We would like to build one wall five stories high and about ninety feet long, tear down possibly three hundred feet of 5-story wall." After nearly fifteen years as treasurer, he confronted a level of deterioration which offended his sensibilities and priorities (such as safety) but on which he had been able to have little impact. Through the ensuing decade, his modest recommendations were either ignored or given only piecemeal attention.[23]

Lockwood-Greene's assessment in 1920 underlined the lack of change and potential result while explaining the low value they placed on the plant:

> We feel that all buildings should not be put in at more than 30 percent of their replacement value today. This may seem a radical depreciation in view of the fact that they will house the machinery and permit of manufacture now but we know that in the near future, in order to manufacture economically and give the help the working conditions *they are entitled* to and *will demand,* at least 70 percent of the replacement value of the buildings will have to be put into them.[24] [emphasis original]

Unpleasant, uneconomical conditions persisted, despite the frequent notice taken of them.

THE CONDUCT OF LABOR

Efforts by management toward more precise definition of individual tasks in the period 1905–30 permit us to make clearer assessments of the activities of the people in the mills. American industries at the turn of the century were in the process of attempting to measure production better, control the time spent at work more effectively, and apply it more efficiently to the tasks at hand. Taylor's popular "scientific management," made famous in his experiments in steel mills and machine shops, utilized concepts long part of the textile mill mentality: cost accounting, production and inventory control, division of labor (specialization), careful measurement of output, piecework or incentive wages, and job analysis intended to enable management to understand and define tasks and workloads. The Boott moved to develop further these patterns (traditional in textiles) through all available avenues, part of the general movement toward heightened control in industry that David Montgomery cites.[25]

The installation of a time clock in 1913 epitomized this thinking.[26] No longer would timekeeping be within the purview of supervisors. Instead, precise and centrally located information would be available to payroll and costing personnel. The hours of the work week in 1916 added up to 54 hours of straight time, 6:45 A.M. to 12:00 noon, with an hour off for lunch, then 1:00 to 5:30 P.M.; work ended at noon on Saturday. When a night shift operated, people put in a 60-hour week beginning at 5:30 P.M. Employees ate supper at their machines.[27] The installation of Veeder 3-shift pick counters on the looms went further. These devices recorded every throw of the shuttle across the loom during each shift. Measuring a weaver's production now became easy and precise, as did determination of a loom's downtime, for whatever cause, mechanical or human.[28] Installing these devices was of a piece with the overall goal of productivity through measurement.

Regarding the nature of this work, Copeland offered a backhanded defense of textile skills in 1910 while offering a racist explanation for the absence of blacks from southern mills: "The negro slave was too ignorant, too clumsy, too unreliable to be employed advantageously in a mill equipped with intricate, expensive, and delicately adjusted machinery." Suddenly the assumption of low-skilled work appears less safe, the descriptions of tasks found in job descriptions less simple, perhaps less complete in their ability to convey the requirements to an inexperienced reader. One is reminded of the phrase from the instruction manuals regarding judgments "not reducible to numbers," actions determined by skill and experience, not measurement, associated with numerous textile chores.[29]

In order for us to examine the nature of work at the Boott, we will consider job descriptions and mill practice in two areas: spinning and weaving. These operations represent the diversity and interconnectedness of work in all areas.

Spinning Room Labor

Spinners tending the warp spinning minded ten sides of frames on 24s–50s yarns, 138 spindles per side, 1,380 per spinner. They creeled roving (to be drafted and spun) on bobbins onto skewers at the top of the frame; they were criticized for removing bobbins before they were empty and cutting off the remaining roving, wasting processed material, ruining the bobbins, and endangering themselves. Since they worked by the piece, the spinners cared primarily to maximize production and did not share the mill's concern for the waste.[30]

Doffers worked in pairs, removing bobbins full of yarn from the spindles and replacing them with empty ones and piecing up the ends to resume spinning. On 23s yarn, a doffer tended 18 frames and doffed each one seven times a week, lifting, replacing, and piecing up 34,776 bobbins in that time. Regardless of the yarn weight, all doffers lifted about the same number of bobbins, although they covered widely varying numbers of frames to do so. A chart described the numbers of people involved with spinning in #6, their tasks, and the numbers of workers in each position. It presented the complement of workers assigned to the rooms for all purposes, not only the obvious tasks of spinning and doffing (see table 2).[31]

Twenty-four workers, in addition to spinners, worked in two spinning rooms; an overseer and a supervisor directed their endeavor; hierarchically and monetarily beneath the other workers toiled bobbin cleaners and yarn pullers, removing tail ends of yarn from bobbins ejected from the automatic looms. Their pay varied according to which type of feeler was being utilized, the new Midget feelers presumably signaling the change nearer the end of the yarn supply on a bobbin; thus did technological change affect even the lowest worker on the mill's employment ladder. The roving man received carts of full roving bobbins from carding rooms elsewhere via the elevator and moved through the room piling them on the frames, on a shelf above the creel, and removing the empties in carts filled by the spinners as they creeled the roving. Doffers placed full bobbins of yarn in hand trucks for the trucker to push into the elevator and then to spooling. Other "specialists" swept, scrubbed, and oiled and cleaned the frames.

While the work of the place was spinning, there were more people attending to the spinning than actually doing it. The amount of oiling and cleaning was particularly noticeable. Top rolls and thread boards were cleaned with a brush every two hours, the roller beam twice a day, top clearers picked three times a day, and other parts cleaned daily or weekly. A checklist for inspection listed fifty-eight separate items to be examined.[32]

TABLE 2.
Spinning Room Labor

Mill	Position	Rate ($)	Rate per Week ($)	Duties
	1 Overseer			Has charge of spooling & twisting also.
	1 Supervisor		$36.00	Has charge of Mills #6-4, #6-5, and twisting in #7-5. These are assistant overseers, and watch the help in their rooms.
	4 Front bobbin cleaners	$.0608 per 1000 bobbins	14.00	Clean the filling bobbins.
	2 Back bobbin cleaners	.0304 per 1000 bobbins	14.00	Clean the filling bobbins.
	4 Pulling-off yarn	.14 per 1000 bobs. old feeler, .105 per 1000 new feeler	16.50	Pull yarn off filling bobbins by hand.
6-4	1 Third hand	.506 per hr.	24.89	Does third hand's work in Mill #6-5 also.
"	1 Fixer	.545 " "	26.16	Fixes frames, bands, and oils.
"	1 Roving man	.4006 " "	19.23	Lays up roving, and takes down empties.
"	1 Trucker	.4006 " "	19.23	Takes yarn to Spooling Department of Mill #1-5, and trucks empty bobbins back.
"	3 Sweepers & cleaners	.308 " "	14.78	Sweep floors, clean frames.
"	2 Roll pickers	.3375 " "	16.20	Clean top and bottom rolls and oil them; help in changing.
"	1 Scrubber	.285 " "	11.12	Scrubs in Mill #6-5 also. Works only 39 hours per week.
"	1 Oiler—waste man	.336 " "	16.13	Picks up the waste, cleans overhead, fills fire pails, also sorts the waste and carries it out.
"	2 Doffers		24.00	
6-5	1 Fixer	.545 " "	26.16	Duties same as other fixers
"	1 Roving man	.4006 " "	19.23	" " " " roving men.
"	1 Trucker	.4006 " "	19.23	Trucks yarn to Spooling Department of Mill #1-5, and trucks back empty bobbins.
"	3 Sweepers & cleaners	.308 " "	14.78	Duties same as other sweepers and cleaners
"	2 Roll pickers	.3375 " "	16.20	" " " " roll pickers.
"	1 Shaftman—oiler	.336 " "	16.13	Picks up waste, sorts and weighs it, then carries it out, also fills fire pails and brushes down overhead.
"	2 Doffers		22.00	

Source: "Textile Development Corporation Report," 1928, box 45, Flather Collection.

Weave Room Labor

The number of looms tended defined a weaver's assignment. The type of loom run and cloth produced affected these loads. Weavers tended 18 looms on moleskin, from 6 to 12 on scrims, 20 to 22 on velvet, and between 6 and 20 on duck, the major discrepancies within a fabric representing old-fashioned versus automatic looms. Battery hands, replenishing the bobbins in the Draper looms' magazines, served between 22 and 65 looms, depending on the speed with which the weft was used up (a product of the loom's speed and the weight [or size] of the yarn).[33]

But the new looms did not define jobs, and equipment could not dictate conditions to the overall operation. In 1928, the TDCR indicated an overload on weavers: "The stops per loom per hour as shown by the actual tests taken during the survey would indicate clearly that under the present conditions the weavers have too much work to insure a high production, and good quality." Poor conditions and poor maintenance blocked full benefit of the technology. They prevented efficiency on the part of worker and machine alike. Typically, these reports anticipated increased production through improved conditions, "allow[ing] a weaver to run more looms with less work than they are doing at present."[34]

The Valentine Report had expanded on the relationship between assignments and "stops" in determining efficiency. Excessive loads were being assigned, a practice which actually reduced efficiency:

> One of the overseers, for instance, stated that English speaking weavers are very reluctant to work on Draper automatic looms where they are asked to tend as high as twenty looms. They flock to Lowell looms where one person tends six machines. That is, in his opinion, the most intelligent weavers consider tending Draper looms too hard. Clearly the capital investment here is linked up closely with human questions such as source of labor supply, selection, training, fatigue, physical working conditions, speed of work, etc. The existing attempt to work out the problem contains too much guess work.[35]

The "many variables" involved indicated the need for objective analysis of conditions in order to properly apportion labor to technology in the mill and ensure efficient use of machinery (regardless of the equation of intelligence with speaking English). Engineers stressed that proper conditions, and loads, permitted increased assignments (stretch-outs) without added effort. Labor tended to disagree and resist, for reasons associated with its different outlook, workplace traditions, and knowledge of which aspects of the work made it more or less difficult, satisfying, stressful, and remunerative. Thus those workers presumably most able to avoid the Draper "automatics" did so, obviously dismissing claims that the loom's higher assignments did not represent greater strain. This conduct flies in the face of the belief that "improved" technology bore the burden of the new loads.

Weavers on all looms repaired broken warp or weft threads, made minor adjustments, and signalled for a fixer for more serious problems. On old-style looms they replenished spent bobbins in the shuttle. Weavers on the automatic looms responded to a loom stop, indicated by one of the stop motions (rather than detecting it themselves), repaired the problem, and restarted the loom. In some instances weavers served as their own battery hands, some "picked out," or corrected, errors and doffed the cloth made. Regardless of the type of loom operated, they had to guard against many of the causes of seconds, or imperfect cloth. Listing a few of the defects indicates the potential for error and the number of sources from which seconds flowed:

> A. Filling Bunches
>
> Many of the filling bunches were caused from the dirty condition of the loom. Lint had accumulated around the filling forks, racks, and shuttle boxes; the batteries were dirty also.
>
> When the shuttle leaves the shuttle box, the filling slackens somewhat, and jerks up dirty lint, which causes bunches. . . .
>
> D. Mispicks
>
> Many of the mispicks are caused by the weaver not finding the pick properly, when the filling breaks.
>
> E. Thin Places
>
> Many of the thin places are caused by faulty loom fixing. When rocker shafts and boxes are badly worn, it is very difficult to set a loom so it will change filling correctly; this causes thin places. The crank arms were found loose on many of the looms; this is a defect that causes the lay to run unsteady. When the lay runs unsteady, the filling fork will not stop the loom as it should when the filling breaks. . . .
>
> I. Oily Spots
>
> Many of the oily spots are caused by oil dropping from the hanger, or shafting. The loom cleaners make considerable oily places, when cleaning the looms. . . .
>
> O. Wrong Draws
>
> Generally the wrong draws are made by the weavers, and they should be held responsible for this defect.

These five problems indicate the variety in a much longer list which defined defects and assigned the fault for different types, some of which derived from the card and spinning rooms. Here again, the interactions between departments of the mill, people in the departments, people and the machines, and even the structure itself (as when floors leaked scrub water onto cloth or shaft hangers dripped oil) came to the fore: direction, conditions, equipment, and workers combined to produce defects. The extent of this list is impressive not simply because of its size, but also because defective cuts, or pieces, of cloth were incredibly common, in one three-day test ranging from 24 percent of the corduroy, 39 percent of the moleskin, 36 percent of the scrim, and an astounding 80 percent of the velveteen.[36] Such a performance makes it clear that no part of the operation, no group within it, achieved the levels of production, whether in terms of efficiency, wages, or profit, that a mill expected at this (or any) time. Frustrated workers led to carelessly tended machines, indifference to detail, mix-ups of different types of yarn, errors in spinning leading to difficulties in the warping; over and over the same types of mistakes took an enormous toll on the work and the workers:

> Slovenly cleaning leaves dirt which forms into webs. These get stirred up in the air and fall on the roving, yarn or cloth. Poor oiling causes dust to accumulate. Feelers in Draper looms if poorly adjusted cause much waste of yarn on bobbins. Careless handling on greasy floors in the weave room makes grease spots on rolls of cloth. Oil cans put in waste boxes makes oily waste. Work by untrained operatives is slower, spoils more, is less adaptable, causes more machine repairs, causes interference with transportation and lowers production.[37]

In nearly every case, mistakes negated the effect of labor previously expended on the cotton, and in many instances they persisted and damaged further work before they were discovered or because they could not be overcome. When things went badly because of these failings, everyone suffered, demoralization set in, wages were lost on piecework, and a vicious circle continued.

The new Drapers and the continual application of improvements to them significantly affected the work of the loom-fixers, the other people primarily involved in weave room work. George Draper claimed they required "fewer calls for the fixer on these new mechanisms." Improved brakes, bearings, bushings and such made the automatics more solid, less prone to adjustments by "shimming."[38] Specific improvements made the loom a stronger, more precise machine, but simultaneously speeds increased and its automatic functions grew more numerous and more complex. As the number of looms assignable to a weaver increased, so did those per fixer. Even small improvements such as the new warp stop-motion, the Stafford thread cutter (which clipped off the trailing end of weft when the bobbin was changed), and the Midget Feeler (which sensed the amount of yarn remaining on a bobbin, triggering a change) adopted at the Boott during this time altered operating circumstances for weavers and fixers. The speed and complexity of the new looms required more precise adjustment or repair. They were significantly more demanding to work on.[39]

Given the innumerable variables in operating conditions involved, in addition to the steadily shifting technology, fixer workloads were a continual area of disagreement between these skilled men and management. In the weave rooms a loom fixer took care of ninety-eight Draper looms or eighty-nine Lowell looms in one case, seventy-eight Drapers in another.[40]

Given the obvious skill and importance of the fixers, the industry's efforts to make their jobs more measurable, more susceptible to management analysis, understanding, and, therefore, control and assignment, were predictable. Despite management's effort to make each job in a textile mill as simple as possible, these workers still dealt with a wide range of problems, the difficulty of which stemmed from the fact that the loom had some seven separate but interconnected motions, each one of which had to be not only correctly adjusted but also timed in accordance with its relationship to the other six. Altering one aspect of the loom's operation affected its interaction with all the others.[41]

The diversity of tasks, further complicated by varieties of looms and cloth, made the elimination of skill difficult, and the fixers' performance remained a major factor in this significant and labor-intensive portion of a mill's operation. The condition of the looms and workspaces directly affected their accomplishments.

Loom fixers at the Boott did not receive high ratings from the consultants who visited. In 1928, they were faulted for only reacting to loom malfunctions, rather than inspecting and maintaining the looms on a regular basis. One example from a lengthy inspection schedule suggested the nature of the work:

Parts to be Inspected	How Often to Inspect	What to Look For
Filling feeler	Weekly	Too close setting; not set close enough; knocking out too many bobbins; is it centering slot, and not touching shuttle; general condition.[42]

This one simple element from a list of thirty-eight items illustrates the extent to which the worker had to rely on judgment rather than definition, feel rather than measurement. The number of things which had to be checked, often in several ways, indicated the difficulty of maintenance and repair. The recurrence of such terms as "general condition, general examination, too large, worn enough, freely, too much lost motion" revealed the necessity of experienced eyes, ears, hands, and minds in these jobs. The report blamed the fixers, despite noting that worn looms made their task "very difficult."

Working closely with the weavers and fixers, but in very limited roles, were the cleaners and oilers. The cleaners in the weave rooms kept the machinery free of lint, dust, and other dirt. Each day, they cleaned every loom with compressed air. Whenever a warp ran out, they performed extra cleaning and wiped off oil and grease. They had to use care in "blowing off" or dirt would be pushed onto the cloth or yarn and damage it. The looms also needed regular oiling, daily for some parts, weekly for others.[43] Because of the number of moving parts and the speed at which they turned, hour after hour, lubrication could make the difference between early or late repair and replacement.

The weave room stood as a microcosm of the mill as a whole, with management and workers dependent on one another in complex ways, responsibilities difficult to sort out with finality. Levels of skill were diverse, and efforts to minimize them continual.

Associated Work

Much more work went on routinely at the Boott that left little evidence in production records. A list of "Help for Yard-Mechanical and Power Departments" in 1919 indicates how much associated work there was (see table 3). This list details some of the many craftsmen needed to maintain the buildings, the power equipment, the plumbing and belts; the twelve men to work in the yard itself, the teamsters to handle the horse and wagon work, and a whole series of skilled people that had to be continuously available for the mills. These male positions paid much better than most factory jobs did. The skills and numbers of people involved were more than those for a New England village.

The Condition of Labor

Consultants found management particularly insensitive to working conditions. In 1911 Hackett wrote:

> In almost all the rooms of the Boott Mills, the common practice was to hang the outer garments on a nail or hook, nearby the machine on which they were engaged. The odors arising from these clothes, especially of a damp morning is considerable, if allowed to hang in the work room unprotected, and it is anything but sanitary.
>
> Factors like the above with those of light, heat and washing facilities, all

> have their influence toward elevating or depressing the attractiveness of employment at the mill, and so the better operative relegates the dingy shops to their distant foreign relations.[44]

Advisers continually related the general work experience to the quality (particularly training) of workers who would be attracted to and held by the Boott.

Five years after Hackett's report, the Valentine Report extended these complaints. It noted that the ramps connecting the floors of the several mills and their different levels were oil-soaked, slippery, and "offer an excellent chance for a fall." Warp beams, empty and full, scattered about the floors of the weave rooms, and particularly those left in unlighted entrance-ways, also invited falls. Windows in such poor condition that panes fell out, particularly when they were opened or shut, posed a danger and led to frequent cuts. The knives worn by workers in the spooling and spinning rooms for cutting away yarn and roving from the bobbins also produced numerous such injuries, and the old and scarred flooring put splinters into the feet of those who went barefoot, a tradition (born of heat and fear of sparks from hobnails) among the mulespinners. These conditions represented more than the apparent sum of their parts: "Habituation to dangerous surroundings dulls the natural spirit of caution. . . . The fact that the insurance company under the state accident compensation law takes care of the financial liability insured has tended, we fear, to lessen the care taken by the management in this respect." In a six-month period in 1916, one man died, 1,029 days of work were lost to accidents, and many other injuries occurred. "When people are living on a narrow margin even though protected by accident insurance, such risks have a more serious aspect than they would under other circumstances." As a result, the better employees sought work elsewhere.[45] The report indicated some of the dangers and implied that they should have been still more obvious and of more urgency to those habitually in the mill. That those dangers awaited a consultant's visit to receive attention carried unsettling implications about management's priorities. The death and the accident toll implicated the operation as one significantly more dangerous than were others at this time. The suggestion that state regulations, initiated in response to overly exploitive conduct on the part of industry, might have led to reduced concern for worker safety carried a heavy irony.

Poor lighting increased the hazards which heightened danger and lowered spirits. Dark interior walls minimized the light admitted by the small windows. Machines ran lengthwise in the rooms, blocking natural light. The windows' once-yearly washing did not adequately address the problem, and "light reflectors are rarely cleaned." Artificial lighting im-

Table 3.
Associated Jobs and Wages in the Yard-Mechanical and Power Departments, 1919

Operatives		Hourly Wage ($)	Weekly Wage ($)	Weekly Hours
1	Second hand	.6325	30.36	
13	Carpenters	.35–.5802	16.80–27.85	
4	Painters	.4771–.5339	22.90–25.63	
1	Mason	.703	33.75	
1	Helper	.4483	21.52	
1	Sweeper	.3306	15.87	
1	Supply man	.4439	21.31	
1	Second hand	.6325	30.36	
12	Machinists	.3407–.6162	16.35–29.58	
1	Sweeper	.3833	18.40	
3	Pipers	.4483–.5964	21.52–28.63	
1	Helper	.3646	17.50	
1	Blacksmith	.667	32.02	
1	Helper	.4194	20.13	
3	Beltmen	.4194–.5280	20.13–25.34	
1	Tinsmith	.5923	28.40	
1	Helper	.5384	25.84	
7	Elevator men	.3728	17.88	
1	Oiler-Night	.3726	19.28	51 3/4
5	Oilers	.3726–.4485	17.88–21.53	
2	Wheelmen	.4612	23.41	50 3/4
1	"	.4612	22.14	48
1	Watchman	.3664	30.78	84
6	" Night	.3281–.3418	27.71–28.71	
2	Draft men			
	1@	.5511	26.45	
	1@	Flat	25.00	
6	Electricians	.4801–.6924	25.69–37.04	53 3/4
1	Electrician	.4762	24.17	50 3/4
1	Welder	.64	30.72	
1	Mechanical helpers	.4483	21.52	
1	Water wheel tender N.	.461	26.05	52 1/2
1	Second hand	Flat	22.77	
12	Yardmen	.3780	18.14	
2	Teamsters	.3729–.4681	17.90–22.47	
1	Assistant Engineer	.6505	38.55	59 1/4
1	Oiler	.461	27.26	"
			40.00	66
1	Fireman	.553	36.49	66
1	Fireman Night	"	"	"
1	Coal passer	.378	24.94	"
1	" " Night	.378	"	"

Source: "Schedule of Numbers of People Employed With Rates Per Hour and Schedule of Piece Rate Prices Corrected to Sept. 29, 1919," p. 37, box 51, Flather Collection.

properly applied produced "a combination of glare and shadows that is very trying to the eye."[46] Quality and speed of work suffered.

No artificial ventilation system brought fresh air into the low studded rooms. Cotton dust and "fly" permeated the air. Now its connection to the debilitating, even deadly, byssinosis, or "brown lung," has been amply demonstrated. The fire insurance company observed that its inspection of the plant left it "very much disturbed at the large amount of fly in evidence in most rooms, and the general [lack of] order and neatness of the plant as a whole." The cleaning force had been cut too extremely, producing a situation conducive to the spread of fire and leading the insurance company to desire "their lines reduced." Worst, they had found "no other mill as bad." To be the worst in an industry noted for such problems certainly gives pause, as does the notation made just a day later that the mill had been cleaned, as if the problem indicated could have been overcome in a day.[47] Obviously the advice given was not being taken seriously.

The antiquated steam heating system did little to improve conditions: "Room No. 6-1 is reported cold in winter, the thermometer sometimes reading 45°. . . . All spinning rooms tend to become too warm. At the east end of No. 6-1 weave room there is a blank wall. Without windows it is dark and hot. The second hand finds it difficult to keep weavers there."[48] This could have been neither surprising nor new, as these problems had persisted for many years. As poor humidification combined with fluctuations in temperature, respiration became congested, and mucous membranes became increasingly sensitive to the dust and dirt in the air. Workers became "more liable to colds, catarrh and grippe infections. If these conditions are prolonged the appetite diminishes. The result is undoubtedly a decrease in vitality and energy of the workers and a substantial amount of lost time from illness." Proper conditions facilitated processing, generated more work, and attracted and retained the best workers: "With improved ventilation, less dust and well controlled heat and humidity the workers would be much happier, more comfortable and would turn off much more and better work. The improvement would more than pay for the maintenance, depreciation, increased taxes and interest on the cost of a better system of air control."[49]

The Boott managers refused to implement the alterations suggested. Needs of a still more basic nature continued to be ignored, as well: "At present employees make rather extensive changes of clothing in the work rooms. Human decency requires better accommodations than this." Furthermore, "the temptation to stealing is too great."[50] An employee had every reason to expect more respectful treatment in a mill of this period.

Well before 1916, when the Valentine Report was made, many companies had become aware of the utility of providing amenities for their workers. Many new work places, including textile mills, calculated the advantage of giving consideration of employee comfort through their attention to lighting, ventilation, lockers, cleanliness, pleasant toilet facilities, and much more; older factories, particularly of this scale, better met standards of decency than did the Boott.

Seventeen years after Hackett's comments, twelve after the Valentine's, the TDCR in 1928 pointed out that unchanged conditions violated laws: "Industrial legislation in this Commonwealth represents minimum community standards below which we may safely say it is poor business policy to operate."[51] The Boott ignored humidity laws, child labor laws, and changing room, toilet, and other regulations. It did not require a "humanist" consultant (such as Valentine) to condemn these practices.

This negligence, and its consequences, were also apparent in the area of cleaning and caring for operating conditions in the mill. With smoking in the mills expressly forbidden, habit and the dust (and accompanying dry throats) led to very widespread tobacco-chewing, creating further health problems: "The lack of cuspidors increases the number of bacteria stirred up in the air from dried floor dust."[52] Valentine laid out the resulting chain of circumstances:

> The importance of cleaning is greatly underestimated. Dust and dirt causes machine bearings to wear out more rapidly, results in an increased use of oil, increases the fire risk, makes more bad yarn and imperfections in the cloth, increases the chances of staining and spotting yarn and cloth, increases the chances for disease and consequent hardships and loss of time, drives away the best workers, makes those who remain uncomfortable and disgusted with their jobs, makes it necessary to pay them more as a recompense for the disagreeable parts of their jobs. . . .
>
> There are no cuspidors near the looms and loom fixers have to kneel down,—often lie down on the dirty floors in order to make the necessary changes in the still dirtier machines. The superintendent complains that the loom fixers shirk their jobs and want more money![53]

It seems more surprising they could find fixers than that those fixers sought raises. A deteriorating building made cleaning more difficult, which led directly and indirectly to waste of labor health, yarn, cloth, and machines; such waste cost money. None of these factors led to efforts at correction. Passages such as those of the consultant, above, are particularly important in keeping before the modern examiner of industrial history the implications of the fact that most of the extant record was created by the managers and owners. The filth in which the loom-fixers worked never appeared in their descriptions or accounts, while the fixers' "intransigence" and "uncooperativeness" recurred steadily.

As the mill's machinery ran faster during this period, noise and vibration increased, particularly in the weave rooms, exacerbating the environmental problems. Conditions there were literally deafening, a cacophony causing hearing damage to workers, as well as being "irritating and fatiguing." While the earmuffs of modern weave rooms were not available, recommended rest periods, which were not permitted, could have eased conditions noticeably.[54]

While Flather belittled the effects of vibration, Valentine did not:

> Vibration is a factor to be considered in respect to the well being of workers as well as machines. Floor vibration is excessive in rooms 2-2, 2-3, 5-3 and 6-5. It causes power losses, bad adjustment of shafting, increased repairs, liability to maladjustment in machines, probable crystallization of metal with decrease in life of the machines, and lowered quality and quantity of work. It causes discomfort, fatigue and occasional timidity among the operatives. It may have a bad effect on certain organs of the body similar to the kidney affections among street car motormen and conductors. These results tend to increase labor turnover.

The consultant specifically noted the difficulty in working in the rear office room, of which the agent Holgate had complained.[55] Valentine called not for innovative treatments, but for those known to be needed and being provided by other companies in and beyond textiles during the 1910s.

While the original Lowell mills had gone to great lengths to separate themselves from the image of English factories in order to draw workers, the Boott had indeed become a dark and dangerous place to work. The conditions reflected badly on owner intent, hampered and discouraged worker effort, and generally hindered the mill's operation.

Chapter Eight

Participants and Prospects

Supervision of the Operation

SUPERVISORS worked in the forefront of the interaction between capital and labor, employees themselves but implementing the point of view of those in charge. They reached their positions through seniority and knowledge, the product of experience and tradition. They instituted policies originating in the treasurer's office and the Counting House. When the higher-ups erred, the overseers bore the brunt of the mistakes' impact on worker-management relations.

The overseers had to cope with the effects of loose management, the lack of uniformity in overall procedure. Despite their best efforts, the mill's practices spoke louder than their words (or its bells, which were drowned out by loom noise). Bad conditions begat bad habits:

> If workers find that lax management puts obstacles in the way of regular work in the form of poor transportation and all the other difficulties, if the company's disposal of waste is careless, if work is not ready for workers when they arrive in the morning, if the work day with the addition of overtime is too long to make it possible to sustain effort throughout the working period—it is impossible to expect workers to maintain a rigid system in all these matters.[1]

The poor quality of direction vitiated the effect of labor's efforts. While some workers came with experience, others had none, and the mill's conditions did not tend to draw the more knowledgeable. Poorly trained at the Boott, they produced disproportionate amounts of waste and seconds. The overseers must have felt at times that they were at the center of this circle, helpless to affect its course. Because of the many uncorrected problems and the poor attitude on the part of management revealed by those problems,

> tightening of discipline is not the remedy at present. Until physical working conditions and wages are improved and greater managerial efficiency is introduced an attempt to enforce strict discipline in wholesale fashion would be apt to drive away many workers—more than the mill could afford

> to lose. Until workers surely know good methods from bad methods, it is unfair to treat them as if the mistakes were intentional. One of the weaving overseers stated, for instance, that the weavers feel that under present conditions they have done their best and that merely showing them the seconds they make is sufficient to cause them to walk out. Under the circumstances they resent a call down.[2]

The workers reacted to the conditions they experienced, the environment in which they were expected to play their roles. They did not feel they could be called to account for errors they could not prevent. The weaving overseer was advised to await a general improvement before he could expect better of labor.

The situation had so constrained the overseers' power that they did not believe they could justifiably dismiss an employee in any but the most extreme situations: They cited "drunkenness, stealing, insubordination, flagrant vice, fighting, physical violence, continued quarrelsomeness and cheating" as the only sufficient causes for firing a worker.[3] Reflection upon these terms gives rise to a picture of a mill with conditions of general unpleasantness punctuated by incidents of a truly abhorrent character. The difficult position of the overseer stood as a metaphor for all workers' situations.

In 1907 Sheldon described the burden the mill's layout and condition placed on supervisory personnel:

> It takes a great deal better talent to run a mill like yours with these handicaps to overcome than a modern mill. It means constant and close scrutiny everywhere all the time, and this makes it hard work, even for a man who is especially adapted for this kind of a mill. In other words, I think it is more difficult to procure the kind of talent that is needed for this kind of a mill than it is for a modern mill, and I think there are few men who would like to undertake the job, because it means hard work, and even then not the best of results.[4]

These cautions applied to the workforce as well. The conflict between labor and management at the Boott went beyond traditional issues such as wages, hours, and workloads to the most fundamental issue: providing conditions in which workers could make cloth. But given such a warning, did management provide the "close scrutiny" and planning called for? For nearly ten years after the takeover, Wellington, Sears ran the mill to suit its fancy, to facilitate its sales, without regard to the internal problems inherent in the operation which their policies of short runs of many styles exacerbated. Hackett's report in 1911 suggested little had been done by the "new" owners to improve operating conditions. For example, "The spinning room as a whole did not anywhere but give indications of unfavorable management." Some weave rooms offered similar conditions, where looms twenty-five to thirty years old ran without any modern attachments, lacking even warp stop-motions and temples. On the other hand, in the rooms of Draper looms the equipment was up to date but conditions were not.[5]

Hackett was as free as Sheldon had been (and Valentine would be) in his characterization of the nature of management thinking at the Boott: "For a long time and today the mills have been inclined to belittle the matter of the human element of the employees, as a factor in organization."[6] At the Boott, successful operation, profitable running, did not lead to improved conditions or more attention to the employees' needs as identified by hired observers. Instead, it justified perpetuating current conditions which required "hard work, and even then not the best of results."[7]

WORKFORCE

As early as 1910, Agent E. W. Thomas admitted the importance of considering employee welfare but found it impractical to give any attention to it. At the same time, he asserted that the "class of operatives . . . was drifting from bad to worse, and even the worst were becoming hard to retain." Instead of the American, Irish, or French Canadians of the past, employment "has degenerated so that only the Poles, Italians, Portuguese and Greeks are being attracted." Ironically, he complained that these "degenerate people" recognized the nature of their situation: aging machinery, which required increased effort amidst poor working conditions, offered minimal wages, little opportunity to increase income through harder work, and only slight chances for advancement.[8] They saw the Boott as a poor place to work.

In keeping with the "100 percent Americanism" of the time, attention to the nationality of the Boott's workers appeared repeatedly. Thomas provided the most complete account in a letter to Flather in 1912. His figures, which also divided the employees by sex and occupation, traced the movement of immigrants into the mill and its various departments. They must be used with care, not only because their reliability is unknown while the prejudice which produced them certain, but also because rooms of employment may be misleading as to the quality of jobs held. For example, the Poles and Portuguese in mule spinning may have been found more often in the low-paying jobs of back boy rather than in the more prestigious and independent mule-spinners' slots. Similarly, in the weave rooms one would find weavers of both sex-

TABLE 4.
Summary of Workers at the Boott Mills, 1912, by Nationality and Gender

Nationality	Males[a]	Females[b]	Total	%
American & English	123	114	237	15.30
Irish	85	189	274	17.69
French	238	172	410	26.46
Greek	228	91	319	20.59
Polish	104	76	180	11.62
Portuguese	72	21	93	6.00
Belgian	2	0	2	.13
Italian	4	0	4	.26
Lithuanian	1	0	1	.06
Turkish	13	3	16	1.04
Russian	3	0	3	.19
German	1	1	2	.13
Swedish	6	0	6	.40
Syrian	2	0	2	.13
Total	882	667	1549	100.00

Source: Letter from E. W. Thomas to Frederick A. Flather, 5/16/1912, Flather Collection.

[a]Percent males: 56.94.

[b]Percent females: 43.06.

es, but in all the better jobs, such as fixer, only males. Still, the statistics are of interest because they were compiled, and because of their content (see table 4). Given the mentality of the time, which equated worker ability with Americanism (or native birth), the preponderance of the newly arrived at the Boott indicated they were losing the battle for the workers that they preferred.[9]

Thomas also offered figures broken down by occupation (see table 5). While Americans and English appear in all but one department (often in supervisory and fixing positions), the chart shows clear concentrations of other groups, such as Portuguese in Picking and Carding. Since the breakdown did not extend to the Boott's many rooms, it obscured much of the segregation by nationality. However, the office workers were entirely British and Irish surnamed even in 1917–18, as were the mill's overseers and most of the second hands.

Occupations were also presented by gender (see table 6). Jobs remain remarkably gender-segregated, even more than the chart revealed. While picking was clearly male, carding's male nature was

TABLE 5.
Summary of Workers at the Boott Mills, 1912, by Occupation and Nationality

Department	American & English	Irish	French	Greek	Polish	Portuguese	Belgian	Italian	Lithuanian	Turkish	Russian	German	Swedish	Syrian	Total
Picking	5	1	1	1	5	13	[a]				1				27
Carding	34	28	27	49	61	33		1		4					237
Ring spinning	12	13	98	111	5	1				3	1		2		246
Mule spinning	8	14	6		4	10									42
Spooling	8	9	56	32	5			2							112
Warping	8	8	9												25
Drawing-in	5	10	5	4	1					4					29
Slashing & Beaming	4	4	4	2	4					1					19
Quilling		1	2	6	0										9
Twisting	16	11	27	26	1										81
Winding	2	3													5
Weaving	79	105	150	86	81	28	2	1	1	1	1	2	1		538
Cloth Room	10	60	5		10	2				3			2	2	94
Yard etc.	46	7	20	2	3	6							1		85
Total	237	274	410	319	180	93	2	4	1	16	3	2	6	2	1,549

Source: Letter from E. W. Thomas to Frederick A. Flather, 5/16/1912, Flather Collection
[a]No entry indicates that no one of that nationality held that position.

obscured by the fact that females tended frames in the department. Elsewhere, male overseers, fixers, and truckers obscured the female dominance of the work in such areas as spooling and warping.

Objectively considered, the nationality of the employee could have had some effect on operations: in 1916, about a third of the 2,066 workers were said not to speak English. A partial breakdown by national background at that time listed 637 Greeks, 194 Portuguese, 204 Poles, 370 French, and 30 Turks (Syrians, Lebanese), thus indicating the rapid influx of the newer immigrants to meet wartime production demands and to replace experienced help moving to better jobs. Instructing those who did not speak English, then, would have to have been done by a fellow immigrant, tending to concentrate nationalities by job and room. Furthermore, textile mills in New England generally segregated workers by nationality in order to place another obstacle in the way of their organizing.[10]

TABLE 6.
Summary of Workers at the Boott Mills, 1912, by Occupation and Gender

Department	Males	Females	Total
Picking	27	0	27
Carding	120	117	237
Ring spinning	138	109	247
Mule spinning	40	2	42
Spooling	17	95	112
Warping	4	21	25
Drawing-in	11	18	29
Slashing & Beaming	19	0	19
Quilling	9	0	9
Winding	0	5	5
Twisting	39	42	81
Weaving	324	214	538
Cloth Room	49	45	94
Miscellaneous	85	0	85
Total	882	667[a]	1,549[a]

Source: Letter from E. W. Thomas to Frederick A. Flather, 5/16/1912, Flather Collection

[a]Erroneous totals (should be 668 and 1,550) are in original.

WAGES

In order to depict relations between labor and management most clearly, and to define their positions as discretely as possible, we will treat the intermingled subjects of wages, income, and conflicts separately, despite the occasional redundancy that may cause. Wages, the bottom line for workers, continued to show dramatic changes during this period. In June 1907, wages in carding showed increases from $11.25 to $11.80 per week for a fixer, up fifty cents to $11 for the two grinders, with most of the other eighteen carding workers earning between $5.35 and $6.75 after raises of about $.25.[11] Other departments gained similarly. Weavers' rates displayed great disparity according to style, suggesting the varying degree of difficulty in producing the different fabrics and reflecting the contrast in amounts of cloth produced on the LMS looms and the Draper automatics (highest rates would have been for complex, fine cloth made on one of the slower LMS looms, lowest for the pieces made on a weaver's larger number of Drapers rapidly producing a coarse material).

Eva and Alvana Desroches, weavers, 1903. Courtesy the Lowell Museum.

Polish residents of Boott boarding house, ca. 1911. Courtesy the Museum of American Textile History.

National Guard confronts striking workers, 1912. Courtesy the Barr Collection.

Boys on Locks and Canals fence, which guarded the river, with Boott Mills extending to bridge in background. Courtesy the University of Massachusetts, Lowell.

Yard gang cementing yard over penstock for new hydroelectric turbine plant, 1922. Courtesy the University of Massachusetts, Lowell.

Promotional shot, "Packing Towels," 1928. Courtesy the Philip Chaput Collection.

Fixer working on fixed-flat cards with Wellman self-stripper, installed in 1880s, ca. 1935. Courtesy the Estate of Frederick Flather.

Women inspecting finished cloth, ca. 1935. Courtesy the Philip Chaput Collection.

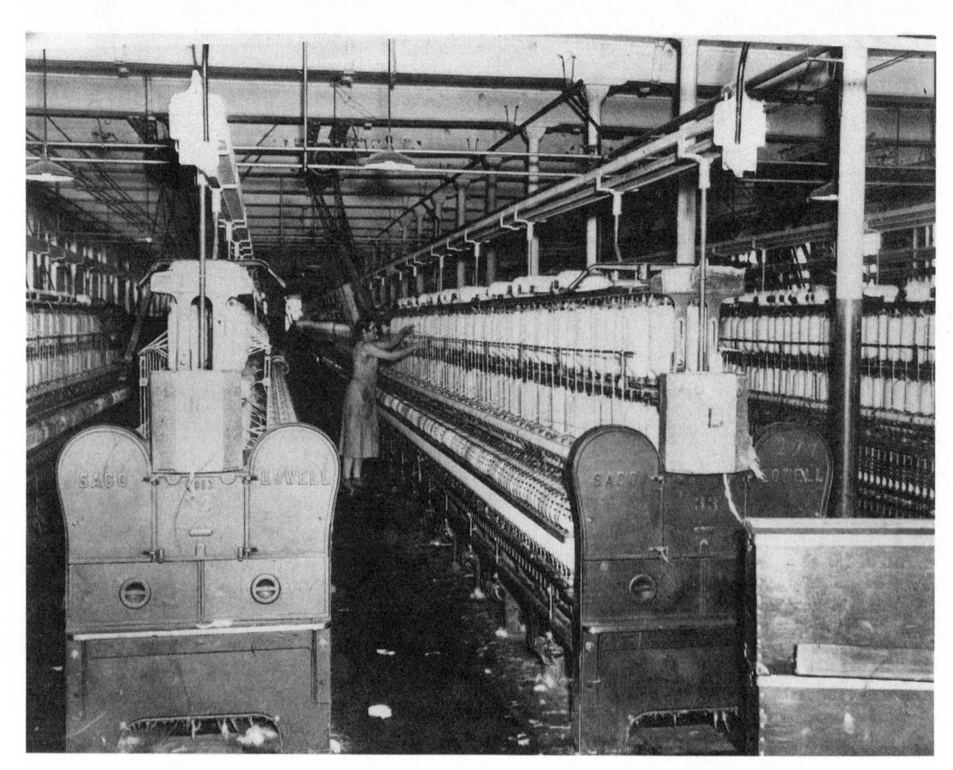

Woman tending ring spinning, 1938. Courtesy the Lowell National Historical Park collection.

Promotional booth at the Boott, World War II. Courtesy the Philip Chaput Collection.

Striking workers picket, 1940s. Courtesy the Museum of American Textile History.

Quitting time, 1940s. Courtesy the Philip Chaput Collection.

Less than a year later, on March 30, 1908, wages fell to levels lower than before the 1907 increase. Reductions of more than 10 percent appeared throughout the mill, and the city, as the workers absorbed the effect of yet another economic downturn. By September 1910, wages still had not recovered their early 1907 level.[12] Labor supply operated against other currents in determining wages:

> Textile manufacturers' access to seemingly inexhaustible supplies of labor and their success in beating down the many strikes by French-Canadian, Polish, Greek, and Italian operatives between 1903 and 1906 kept the average weekly wages of operatives in Lowell, Lawrence, and Fall River no higher in 1911 than they had been in 1893, despite the increases in both productivity and cost of living during the intervening years.[13]

Increases came slowly, and significant improvement had to await war in Europe. By 1916 war work was plentiful (particularly on duck for the military), workers scarce, and wages advanced twice, 10 percent in January and 8 percent in the spring. Because of the narrower range of goods involved and increased production, the first wage increase was overcome, or canceled out in terms of cost, quickly. In management's terms, the wage increase had not cost them anything. Had they been able to attract more help, they could have sold further production. Because of the increases for hourly workers, 2 1/2 percent of profits were appropriated for distribution to about fifty employees on salary who had not gotten the wage advances. Five thousand dollars was distributed in amounts of $15 to $500, with the latter presumably going to Agent E. W. Thomas, who was making nearly $135 a week, $7,000 a year, in marked contrast to hourly workers earning between $6 and $35 a week.[14]

A 1916 report broke down earnings in another manner and indicated the number of people, male and female, who worked at the Boott (see table 7). The lowest-paid group of males appeared to be boys ("males day"), who were unable by law to work at night, as were the women, the only others to earn so little. Women's wages showed marked concentration in the lower groups. Protective legislation written in age- and sex-specific ways thus limited the earning capacity of those it aimed to protect. Even among men, few earned more than $15 a week. Overtime work added about one-third to the pay of some adult males, particularly those in carding and weaving, for four hours' work a night, five nights a week (an enormous demand on their energies).[15] The differences in earnings between those at the lower end of the scale and the overseers, foreman, and second hands were accentuated by the latter's more steady work, overtime, and bonuses. Skill, access to jobs, independence from the moment-to-moment flow of materials, and legislation all perpetuated the male advantage.

TABLE 7.
Workers at the Boott Mills, 1916, by Wage and Gender

Earnings	Number of Workers			
	Males/day	Males/night	Females	Total
$6.00–$ 7.99	103	—	194	297
$8.00–$11.99	487	167	373	1,027
$12.00–$14.99	231	112	65	408
$15.00–$17.99	155	62	7	224
$18.00–$35.00	36	35	—	71
Total	1,012	376	639	2,027

Source: Valentine, *Audit,* p. 168.

The problem of maintaining the separation of the male salaried elite, who did not share in the hourly raises, from the other men led to a rare recorded disagreement between the treasurer and his board. Flather gave discretionary bonuses of $25 to $100 to various personnel, for example when certain production goals had been reached. In one instance, when "a man made a supreme effort in straightening out a strike based on ignorance and misunderstanding and I gave him $100."[16] Doling out the money this way gave Flather personal credit for the largesse and enabled him to reward individuals rather than classes of employees. In 1917, however, the directors authorized a bonus of three times the percentage of dividends times their annual pay for salaried employees. This profit-sharing plan was repeated in 1918, but not the next three years, and resumed in late 1921, when $3,447 was divided among nineteen employees, including Flather. The supervisors' raises insured that they were paid at least 10 percent more than previously and also at least 10 percent more than those they supervised.[17]

The May 14, 1917, wage increase had represented the culmination of a series of increases beginning in 1912, which, according to Flather, "raised wages 10 plus 10 plus 8 plus 10 plus 10 percent" (from 1912 to 1917). At the same time they had eliminated overtime and other "subterfuges" previously used to pay more than the local scale in areas where "labor difficulties" had made it necessary.[18] The directors voted raises for supervisors, to match the recent wage advance, and bonuses for department heads "by reason of the company's exceeding goal established by a Director; . . . [and] to change conditional portion of department heads' salaries into part fixed and part conditional," based on the production of those they supervised.[19] The practice of paying, in effect, piece-rates to supervisory personnel originated in the 1840s in Lowell and marked one of the major steps in the increasing exploitation of the textile worker for which the 1840s and 1850s period was infamous. Basing overseers' income on the production of the people they supervised separated their interests from the workers' and increased significantly the pressure on labor to maximize speed. To reinstitute this approach of conditional wages for overseers indicated a continuing belief on the directors' part that not even supervisory personnel could be expected to exert themselves sufficiently for their employer's benefit without special inducement. The board perceived a separation of its interests from those of even its most elevated employees (including the treasurer) which it hoped production bonuses would overcome.

A "Schedule of Number of People Employed with Rates Per Hour," along with a "Schedule of Piece Rate Prices for September, 1919," give a good indication of the Boott's situation at the end of World War I. They reveal a new 48-hour week and a substantial increase in personnel. The rise in numbers reflected the lesser ability of some of the help available in wartime, although added fixers could have been a response to aging equipment run hard to take advantage of government orders. The situation in weaving was typical. Loom speeds had not increased, and work loads were comparatively low, a result of the lack of skilled weavers and the close inspection of government work. Fifteen fabrics represented the product of the fine goods mill. Increases in numbers of fixers, cleaners, and oilers indicated a redistribution of available skills to keep things going, despite the disruption of standard job distribution such changes represented and the problems that they posed for the future.[20]

Intense competition for labor during the war, steady inflation, and a flow of government contracts necessitated two 15-percent wage increases in 1920.[21] Later that same year, as the national economy once again struggled, New England directors met in Boston and agreed to implement a 22 1/2-percent *reduction* in pay across the board along with a 25-percent cut in output, and thus wages reverted to the November 1919 level for those who remained at work.[22]

In January 1922, a 20-percent cut in wages was imposed at many Massachusetts mills, but not at the Boott. A widespread strike overcame that cut where it was attempted. In April 1923, all the Lowell mills agreed to a 12 1/2-percent wage increase. When the Lowell Scale was reduced by 10 percent in 1924, the Boott again did not join, an unusual deviation from local practice.[23] By the end of 1924 weekly rates for spinners were projected as $21.63, doffers $25.42, weavers $22.42–$25.20, and loom-fixers $32.81.[24] These rates would result from full weeks with sufficient production. In 1925 time and a quarter for overtime appeared. Wage cuts in the twenties came at a time of rising production, an anomaly which led the American Federation of Labor (AFL) to adopt a new policy of seeking wage increases in proportion to increases in production. In addition to the cuts, the office was "working to put whole mill on piece rates," another tactic to increase production while reducing pay.[25] Piece-rates, long the standard in all measurable jobs in Lowell (weekly wages simply reflect average piece-rate earnings for most employees), did more than attempt to inspire effort. As Scranton points out, they formed part of an overall strategy: "Alternating busy and slack seasons shifted the wage or employment risk onto workers; piece-rate schedules made production expense dependent on the goods made, not on the length of the workday [or week, or month] or the reliability of the machinery." Piece-rates thus further insulated the investors from the factors associated with running deteriorating equipment amidst difficult conditions.[26]

Flather's salary, $7,500 per year in 1905, rose to $10,000 in 1910, $12,000 in 1919, $15,000 at the end of that year, and $20,000, or nearly $400 per week, in 1925, in addition to which he received the annual bonus of about $1,500. His stockholdings in the mill made the contrast between his income from it and that of other employees all the more striking. He kept two Cadillacs during this period.[27]

INCOME AND THE COST OF LIVING

In addition to the rise and fall of wages which occurred at all the Lowell mills, several other factors must be considered in evaluating the pay of the Boott Mill workers: availability of work, cost of living, and type of pay. Textile cities had high unemployment even in good years and "the most severe unemployment problems in depressions." In the period 1900–1920, about 20 percent of the workforce was laid off each year for an average of three months.[28] The ease with which a cotton mill could stop and start production combined with the availability of experienced workers to facilitate efforts to make employees bear the brunt of fluctuations in any factor affecting the industry. In 1907, for example, Flather was informed that a decision had been reached to curtail production by 25 percent at the Lowell mills for two months. A slowdown in 1908 led to a 20-percent drop in production, with a parallel effect on employment.[29] The Boott operated on a three-day week for a time in 1920, closed for a period in 1921, and in the spring of 1924 ran one or two days a week "while business was dull."[30] Overall operation stood at 50.8 percent in 1921, 54.6 percent in 1922, and 78.9 percent in 1923.[31] Textiles remained very much part of a boom or bust economy.[32]

No unemployment insurance compensated Boott workers when business trends made it more economical to suspend than to operate. Savings or earnings from family members working elsewhere had to be relied upon. John Rogers Flather noted it was sad to lay people off, inspiring to bring them back.[33]

Living costs rose dramatically during the war years, and contemporary estimates offer intriguing views of the ability of wage increases to keep pace. For a couple with three children, Flather believed the minimum cost of living to be $1,000 a year; Agent Thomas guessed $600; both the superintendent and Mrs. Gilman, who had been working in the office for twenty-five years and had some familiarity with the workers' home situations, suggested $750. In 1911, Scott Nearing's *Financing a Wage Earner's Family* assumed that an income of $750–$1,000 was needed for a decent living, whereas U.S. government estimates called for $691–$732 minimums in 1911, depending on the *nationality* of the families! In 1913 the average annual income of cotton mill employees in Lowell was $454, middling compared to cotton workers in other Massachusetts cities, but well below the $625 of boot and shoe workers or the $692 of machine-shop employees, and far from the amounts needed for subsistence for a family. Since individual employees at the Boott did not, in fact, earn enough to support one person, let alone a family, the Valentine Report noted that "the mill is making heavy drafts on the physical reserves of a large proportion of its workers."[34] Wages at this level offered a strong stimulus to the perpetuation of the extended family, since smaller units had still more trouble making ends meet. The distance between incomes and cost estimates, even by management, demonstrates a startling consciousness on management's part of the insufficiency of the pay.

Furthermore, the Boott's position relative to other Lowell textile mills (not to mention other wartime employers such as U.S. Cartridge, a major competitor for local labor) added to its problems: "Since the work is more irregular at the Boott than at several of the other Lowell cotton mills, this mill is at this added disadvantage." Flather noted that the more skilled employees left town during depressions.[35] The mill attracted only the "poorest grade of workers," and none of these workers could "maintain what could seriously be called an American standard of living."[36] During 1928 and 1929 production increased, with two and even three shifts running. However, conditions kept earnings below those of other Lowell mills, to which Boott workers continued to go whenever possible.[37] As the consultants pointed out to Boott management, payment for labor had to be sufficient to maintain its physical and mental well-being throughout the year or the business was insolvent:

> The workers are in reality far more in need of regularly assured income than most capital holders. Workers are able to save little if any out of present wages. They have no property to fall back on. Rents and grocery bills must be met within the week or month. Most unskilled wage earners are not over four weeks from destitution if employment is suspended. . . . These reserves of health, strength and energy constitute the workers' investment precisely as the capital reserves constitute the capital holders "investment."[38]

The Boott workers existed in a tenuous position. Their pay was low and far from assured, and it offered little leeway for the risks of sickness, unemployment, and reduced capacity in later years.[39] The degree of the dependence on an operation which did not work efficiently and which they could not correct or improve through their efforts frightened and frustrated them. Furthermore, and of particular importance in an industry where survival required the earnings of an entire family, the number of jobs available dropped drastically in textiles. Lowell manufacturing employment fell from 40,000 in 1918 to 30,000 the next year, and to 15–16,000 by 1936.[40] The industry's flight and the stretch-out continually eliminated jobs, enlarging the labor pool and depressing wages.

A final note on wages gives us a perspective on the relative position of some of the players. When John Rogers and Frederick Flather came to work in 1923, they at first served without pay, presumably as they were learning the ropes. After that, they began to receive $24 per week, plus the bonus given to salaried workers, equal to the pay of a worker with significant skill and experience. In 1926 they received a 25-percent raise, and their pay began to be sent to them from Boston, rather than from the mill, perhaps to avoid hard feelings on the part of other salaried employees.[41] Flather Senior received, in addition to his salary, varying amounts as director of the Merrimack, the Tremont, and the Suffolk mills, which were all controlled by the same investors as the Boott. The Merrimack paid him $12,000 per year, the Suffolk, $7,500, in apparent continuation of the practice of skimming high salaries for directors from the budget before paying other stockholders (a practice attacked by J. C. Ayer as early as 1863). He also received several tens of thousands of dollars a year from investments. His wife benefited from a $350,000 trust fund set up for her and her two siblings by her father in 1907.[42] They could, and did, buy one another jewelry worth more than one of the workers could make in a year. They enjoyed Cadillacs starting in 1916, trips to Europe lasting four to five weeks, and yearly vacations in Maine and in the West. When the boys married, each received a honeymoon trip worth about $10,000, one to Europe, the other to Egypt.

LABOR'S RESPONSE

While strikes were rare, the numbers of employees who left the Boott reflected the dissatisfaction among workers there. During this period in general, labor's "only way to protest ubiquitously bad conditions was to quit." With few other tools for resistance, quitting became a statement: "During the 1920s, leaving a job was an important way for a worker to respond to oppressive working conditions, given the lack of collective alternatives." The Boott's turnover consistently exceeded that of other textile mills. It averaged 63 percent in 1913–14, despite its being a time of recession, and 126 percent between 1914 and 1918, when a labor shortage made alternatives available.[43] In 1916,

> One weave foreman estimated that he had lost from his room twenty to thirty workers in the last six months out of a normal working force of forty-three. An overseer in charge of spinning and spooling estimated that he had lost two hundred and fifty in the past six months. The overseer formerly in charge of carding estimated his last year's loss at one hundred out of a force of approximately four hundred. Most of the overseers declared that they could not estimate the amount of turnover among workers in their rooms.[44]

Twenty-four of the 36–42 office workers left in 1916, and 41 men at the level of second hand or above quit in the period 1914–16. Three quarters of the weavers leaving in half a year and 250 out of roughly 350 spinners and spoolers indicated fleeing, not quitting. These rates are far higher than the 30–60 percent per year cited by Caroline F. Ware as average. Since the most qualified workers were the ones most able to find work opportunities elsewhere and to exercise choice, the significance of the losses became enormous. For example, the employees in the less desirable carding room jobs were less mobile, but also less important to keep. The numbers of favored office help leaving, and nearly complete turnover of important supervisory personnel, indicated that the Boott was losing the unquestionably skilled employees others moved to retain. Training new workers represented an obvious cost of turnovers, and training new supervisors cost a great deal more than training laborers. These costs included time spent teaching, record-keeping, accidents, errors, production losses, damage to equipment, and idle machinery. Causes of the unusually high turnover observed at the Boott included "irregularity of work, poor physical working conditions, and low wages." Fabric style changes (making earning at piecework a challenge), inadequate transportation of material, and poor coordination between departments aggravated the situation and caused workers to be laid-off temporarily or to lose money on piecework, leading to discontent and relocation. Higher earnings available at other Lowell cotton mills and the much higher pay at the U.S. Cartridge Company encouraged migration. Conditions at the Boott and the workers' inability to affect them led many to vote with their feet.[45]

As always, unorganized workers had little ability to confront management. Unions existed only among the mulespinners, slashers, weavers, loom-fixers, machinists, carpenters, and firemen (tending boilers, not extinguishing fires). The heavy turnover and arrival of inexperienced workers held down the percentages of members in each area, but the mulespinners were perhaps a third union, the slashers and firemen were entirely union, and the loom-fixers the most powerful. Even in these areas of craft unions, labor in Lowell lagged far behind other centers, particularly Fall River and New Bedford, in its degree of organization and strength, for several reasons. The contrast was striking with regard to its nearest neighbor, Lawrence, where labor unrest was far more common. While textiles dominated the city similarly there, divided ownership of the several major textile mills made a united front facing labor impossible. The greater number of highly skilled jobs in woolens and worsteds also required (and attracted) experienced immigrant workers. Many of these brought traditions of organization with them, and some, particularly the Franco-Belgians and Italians, came with highly developed radical political beliefs as well. Their absence in Lowell indicated the success of long-standing Lowell approaches: (1) joint ownership and joint action, (2) production of coarse cottons, and (3) anti–craft union activities.[46]

The craft unions were notoriously slow to organize women, thus ignoring nearly half of the Boott's workers. Prospects for organization were also reduced by the mill's high turnover rate and the preponderance of foreign workers and their diverse languages, factors noted by all observers.[47] Of thirty-two strikes (of all sizes) at any Lowell cotton mill between 1900 and 1914, only one, in 1912, was both successful and involved money. In 1909, when twenty bobbin-boys struck the Boott for a day seeking a 10-percent raise, they were not only unsuccessful but their leader, Thomas Grogan, was fired.[48]

The notable exception took place as an aftermath to the famous Lawrence, Massachusetts strike led by the Industrial Workers of the World (IWW, also known as Wobblies), advocates of "One Big Union" of the working class. The strike began in January 1912, after earnings fell when the law reduced women's and children's hours by two, to fifty-four per week. The strike, with its famous battle cry of "Bread and Roses," caught the attention of people around the world. Lowell papers followed its developments and approved the cause but not the tactics. Lowell police and militia assisted in Lawrence's efforts to prevent labor action.[49]

IWW organizing in Lowell increased dramatically; 1,525 Portuguese workers were said to have joined. Visits from IWW leaders William "Big Bill" Haywood and Elizabeth Gurley Flynn encouraged strike talk in the city. In March 1912, after the strike's successful conclusion, the Lowell mills announced a wage advance said to equalize Lowell and Lawrence standards, but it was readily apparent that they offered significantly less than the 10-percent increase won in Lawrence and elsewhere in New England. The IWW moved first against the management of the Appleton Mills, where a majority of the employees were members of the union. The Appleton closed. Workers then drew enough employees out from the other corporations to lead all the owners to join the lockout on March 26.

The IWW immediately called in Haywood, Flynn, and William E. Trautmann, set up picket lines, and organized a strike committee along ethnic lines, as they had in Lawrence. Demands included a 15-percent wage increase, double pay for overtime, the right of weavers to weigh their own cloth, and reinstatement of all striking workers without discrimination. Employee committees presented these demands to the agent where they worked. E. W. Thomas refused to meet the Boott workers' committee. The agents announced they would instead meet with the crafts, the Mule Spinners', Loom Fixers', and Weavers' Unions said to represent about 25 percent of the workers and affiliated with the United Textile Workers (UTW). This divisive strategy had been attempted in Lawrence, and UTW President John Golden came to Lowell with AFL organizer Carl Wyatt. They contacted their 1903 ally, the Lowell Textile Council, now merged with the Lowell Trades and Labor Council, and together they condemned the IWW while issuing demands similar to the Wobblies'.

IWW members picketed and paraded. On one occasion they marched through the Belvidere section of Lowell, where John Rogers later remembered watching as the family's chauffeur misdirected the workers, sending them away from their target, the Flather home.

Mills in other New England cities had granted the 10-percent raise won in Lawrence, returning pay to the level of 1907. The New England Association of Textile Manufacturers censured Lowell for not joining the increase. On April 13, after Flather met in Boston with representatives of the other Lowell mills, a 10-percent hike was announced for an April 22 reopening. The UTW and the Greek community, which had been allied with the IWW, announced acceptance. The IWW had little choice but to go along, but did so with the reservation that other issues had to be settled between the employees and management of each mill before a return. Shop committees formed for each corporation would act as go-betweens in disputes between management and labor. The confrontation had been successful for the workers, who achieved not only a raise but also, potentially more important, a degree of recognition for their organizations from the owners.[50]

According to the agent in 1916, the small unions had positive effects on the operation. They might have raised members' wages, although the superintendent doubted it, and they reduced the numbers of looms per fixer, but they insisted upon sobriety, demanded strong evidence from a member to initiate a grievance, and steadied the output of the mill. The agent dealt cordially with the unions (though only when they initiated contact), advised them to organize more classes of workers, and knew personally the secretaries of the Loom Fixers' and Weavers' Unions, the latter affiliated with the AFL.[51] The UTW local, an AFL affiliate representing many craft workers, went so far as to expel radicals and claimed a deal with Lowell agents to have them fired, as well.[52] Unimpressed by these advantages, Flather and the other directors continued to resist organization.

In March 1917, Flather reported on a conference involving a Greek interpreter "in view of settling the strike. Gregg [a consultant in the Valentine Company] turned around from the telephone and said 'the Greeks have joined the union' which *I am glad to hear.*"[53] The Greeks had traditionally acted independently of organizations, though often in concert with them, and this incident indicated a simplification of the situation with which Flather had to deal.

When the Boott's machinists threatened to strike during the war, the threat of federal intervention led the mill to settle.[54] Heavy government production on cost-plus contracts meant high profits, enabling the corporation to meet demanded pay increases brought on by a skyrocketing cost of living and higher wages elsewhere.

In June 1917, skilled workers affiliated with the UTW demanded a 15-percent wage increase; 30 percent of the operatives struck on July 1, after the mills agreed to only a 10-percent increase. Increased support from less skilled workers led to a walkout by 70 percent the next day. Soon both sides agreed to arbitration by the State Arbitration Board. It considered cost-of-living increases, which had more or less equalled raises, and the fact that Lowell workers earned less than textile workers in other Massachusetts cities, then granted the 15-percent increase, with the debilitating proviso that strikes were forbidden for the duration of the war.[55] In exchange for the extra 5-percent raise, the union had lost its bargaining chip.

In January 1921, wages were cut 22 1/2 percent in response to deflation from the 1920 recession. After a second, 20-percent cut in 1922, 66,000 workers walked out at sixty-five mills across New England. The UTW led strikes at four Lowell mills,[56] but the Boott avoided the strike by not joining the second wage reduction. The attempt to cut wages was foiled, in Flather's opinion, by the refusal of the giant American Woolen Company and the Fall River and New Bedford mills to participate.[57]

The landscape of the period became littered with the carcasses of failed unions. The UTW, after a growing membership during the war, "suffered crushing blows from employers who established the open shop in spite of militant strikes in New England milltowns." The Amalgamated Textile Workers of America, founded in 1919 and opposed to the UTW, died by 1922, and the National Textile Workers Union, the Communist Party contestant, was "decimated prior to the crash." The collapse was complete.[58]

MANAGEMENT'S RESPONSE TO LABOR

Management attended to the return on investment. It pursued superficial technological "fixes" and administrative readjustments. An episode involving the mill's firemen reveals Flather's attitude toward technology and labor. In 1917, Flather recommended acquiring oil-fired boilers for several reasons, including the following: "To reduce firemen who are now somewhat independent." The Boott was one of only two Lowell plants still using hand-fed coal-fired boilers, and in this instance their lack of investment required them to deal with a small group of organized workers. After a strike by firemen in 1918, Flather again urged the expenditure of $45,000 on a fuel-oil plant. With money for equipment scarce, as always, he wanted to use that amount to combat this pocket of organization within the labor force.[59]

Communication with many of the employees was limited. Signs regarding hours of employment for women and children, liability insurance, and other notices were posted in English only.[60] The large waves of immigrants that arrived before 1914 learned little about cotton mill operations. Management did not undertake to train workers, leaving that to their compatriots. Employees new to the industry, as great numbers were, remained ignorant of the overall process. Valentine observed, "It is unfortunate that a false notion of the slight amount of ability required to operate cotton machinery has been so dominant." He recognized the complexities of jobs long described as "low-skilled." When trained help suddenly became unavailable at the time of World War I, the mill had none to fall back on and its critical importance became apparent. Because of the low wages the Boott offered, it competed for workers at a disadvantage, yielding a shortage of labor, inefficiency, high turnover, and lack of interest in production among those it could hire.[61]

Nationally, a strong movement known as welfare capitalism arose in an attempt to influence workers' attitudes during this period. Plans for employee stock purchases, savings, pensions, and insurance became common. Safety and medical facilities drew attention, as did sports and classes outside working hours. Such efforts aimed to increase production through labor cooperation. They also worked against class consciousness and trade unionism among the workers.[62] At the Boott, one found only faint echoes of such approaches.

Management there made certain efforts that it interpreted as demonstrations of concern for employee well-being. Along with other major corporations, it supported the local hospital where low-cost treatment was available to employees and their families. In 1930, however, the corporations turned the hospital over to the Franco-American community for $1. After that the Grey Nuns (Soeurs Grise de la Croix) of Ottawa ran the hospital and they, rather than the mills, received any credit associated with the operation.

Management style at the Boott ran to the idea of granting unpaid vacations and making the occasional donation of $1,000 to the North American Civic League for Immigrants "to work with Greeks outside the mill." Only in the rarest of instances would they provide anything resembling a pension for a worker, as in 1930 when they allotted $100 a month *for eighteen months* to the widow of Albert Z. Abbott, their cotton classer for twenty of his twenty-three years of employment. Even this high-level job secured only brief benefits.[63]

They also, in the years around 1912, rented Boott Camp, on a pond near Lowell, where a few female employees could go for a week's vacation or a weekend during the summer.

> Some learned to cook and to care for bedrooms in a way which they never knew before. . . . Mothers who work at the mill were allowed to come and bring their children, and many of the English speaking women of the Carding and Weaving departments took advantage of that. . . . We cannot tell how much good has been done through this "fresh air" home as yet, but many women and girls who are steady reliable workers at the Mill were exceedingly sorry when the season ended.[64]

Mrs. Gilman's letter raised questions as fast as it answered them: how had these people cared for bedrooms before, and how had they gotten along without her methods? What did the non-English-speaking women do during their idle hours with their children, and why? Just what "good" was expected, and how did the other workers feel about the season's ending?

Mrs. Gilman also played a role in the Working Girls Club, a Boott Mill organization which in 1911 thanked Flather for a new piano in the Recreation Room: "It will furnish us a great deal of pleasure during the noon hours and can be used for our entertainments in the future."[65] The club, which seems to have had the same officers as the camp, used the Rec Room for annual meetings featuring entertainment, lunch, and remarks by several speakers, including Flather and Thomas. It also had a Christmas party, and put on a play, "Rebecca's Triumph." Surnames suggest its members were largely of English and Irish background. The club must have helped keep alive among some the myth of the famous "Lowell mill girls" who combined work and intellectual or artistic expression. A company baseball team and occasional outings of the office staff completed this minimal social-welfare work.

Management claimed that a number of factors hindered more fundamental contributions to the labor-management relationship. In 1917, Flather blamed high indebtedness resulting from the takeover, infrequent dividends [although biennial payments had been passed only twice since 1905], and "the wide discrepancy between capital and assets" as obstacles to improvements which not even the $716,534 profit of the previous two years, from 1914 to 1916 (before the best war years), could overcome. Flather acknowledged the effects of this kind of thinking:

> The result of all this on the workers has been unfortunate. The management has deferred many improvements in physical working conditions, the desirability of which it was perfectly aware of. Needed improvements in plant and equipment have been postponed. Faulty accounting methods have made knowledge of waste less effective and created a slight feeling of insecurity in the management. Wages of workers, supervisors and office staff have been kept as low as possible. Management, staff and supervisors have all been worried and spent too little time on improvements in methods and forms of organization. The Treasurer has hesitated to enlarge the size and functions of the office staff as he wished to. Slackening of initiative, inefficiency, discontent, and high labor turnover have resulted.

Flather presented a defensive response to the criticisms of the Valentine Report. His response denied any failure to understand the operation's problems, claimed that not only wages, but also administrative salaries were minimal, and that worry prevented action. The cause lay in directors' policy, which allowed the mill to decline as the cash it generated was set aside for dividends and liquid assets. Instead of reinvesting the large quantities of profits (see below) from this period, improvements came from borrowed money, and by 1914 the Boott carried $1,500,000 of indebtedness.[66] Flather placed himself and his sons among those suffering from low pay and insufficient assistance. The worry expressed, the anxiety inherent in this defense to the directors, recalls Dr. Ghering's description of Flather as "nervous and overworked." His self-serving account reveals an effort to appear in control of a situation in which he could not free himself from the burden of Valentine's complaints. Serious problems mounted even as he struggled to increase profits.

In 1917, Flather sought authority to carry out the recommendations of the Valentine Report for "lockers, toilet rooms, wash rooms, dressing rooms, lunch room, filter, sanitation, hospital (first aid room) cost $25,000." At another time a painting project was described to the directors: "A systematic effort is being made by painting the interior of the mill to improve the surroundings and thereby the morale of our people and the quality of our work. This work may be considered also as maintenance for by keeping our property clean and well painted it lessens the tendency toward depreciation and obsolescence."[67] These efforts failed to come to grips with the nature and extent of the difficulties, whether in terms of physical conditions, machinery, wages, or general relations. To expect paint to stand between the antiquated plant and "depreciation and obsolescence" defies comment. Flather's request for $100,000 for employee amenities in 1919 again failed to win approval from the directors, underlining their frustration of his modest efforts. The Boott was not to be part of the movement that made the American factory a "safer, cleaner, more spacious, better lighted, heated, and ventilated place" in the Twenties.[68]

Consultants did not find a great deal of skill in dealing with employees demonstrated by management at any level, and their recommendation for an Employment Office in 1916 quickly found approval. This office, it was hoped, would give employee relations more importance, would know and deal cordially with all sources of labor supply, and would direct hiring and training, transfer, promotion, and discharge. "In order to avoid resentment, jealousy and opposition among the overseers it is planned at the start not to interfere with their securing and hiring workers just as formerly." In 1920, a woman was added to the office to deal with employment issues of her gender. A 1923 letter implied that weaning the overseers of exercising their prerogatives was not proceeding with ease.[69]

With such new offices, management intended "to break the power of the foremen." With his power reduced, he could better be encouraged to attain their desired production. Without his ability to affect hiring, firing, and promotion, the foreman lost the status which could enable him to interfere with management's continuing efforts to install features of scientific management for heightened production. Foremen might be attached to old ways of doing things or, even more dangerous, feel an allegiance to the laborers from whose ranks they themselves had risen, in which case they might resist efforts to speed up and stretch out the workers beyond certain levels. Ironically, in this instance the attempt to increase employer control over the workplace and work process worked against individual incentive and loyalty on the part of supervisors.[70]

John Rogers revealed something of his own developing approach to labor relations. Not long after his arrival at the mill he described an incident: "Secured a bunch of keys for Jim Doherty (5-1-W) which he lost and were found in a bag of waste at the Merrimack Utilization. I was glad to help him out. The smile of gratitude on his face was worth it and showed that the laborers are as human as anyone else."[71] The note indicated the distance between management and labor, as well as his empathy for the worker.

In 1924 John Rogers denied a suggestion the mill was providing "philanthropic work" during a slow period. Any such efforts had calculated intent: "Our custom of continuing employees when their output yields no added return is for the purpose of stimulating their sincere endeavor to carry out our instructions, which, if we are sufficiently wise in giving, should result in bettering the condition of all." Management endeavored to inspire loyalty through a stratagem, rather than through investment. John Rogers' comment also reflected the naivete of the new member of management in thinking that cost-free methods could lead to harmonious relations. Similarly, when a strike hit New Bedford, he reassured his absent father: "I believe there is not much interest in the strike here, although I am doing all I can to see that our people are treated humanly [*sic*] and wisely along the lines we talked about before you left."[72] Spend sentiment, not money, might have been the mill's motto.

In contrast to the slight effort in the direction of welfare capitalism, the Lowell corporations worked aggressively to extend their control over the factories. In an age of mergers designed to benefit the participants through the advantages of scale, professional management, and market control, the Lowell corporations watched the rest of business ape their methods of a hundred years before. Similarly, the "hostile employers' associations" that Bernstein finds emerging in the twenties had been around since the beginning in textiles. Green describes a parallel movement: "Using new financial and political resources, the big corporations renewed their struggle for control of the workplace and, by extension, for domination of the social and political institutions." Here again, Lowell had pioneered the techniques he lists: residences segregated according to jobs, local political influence, vocational education to train skilled and clerical workers, investment in new technologies, and an army of managers and superintendents to install scientific management. Significantly, the Boott put little effort or money into such approaches any more. Ownership's position was strong, its investment minimal, the operation profitable.[73]

The economic power of the Boott's directors persisted. For example, Bemis sat on the board of the First National Bank of Boston (as did directors of many of the leading New England textile companies, including American Woolen, Amoskeag, Pacific Mills, Pepperell, Berkshire, Farr Alpaca, the Knight group, Merrimack, Hamilton Woolen, Nashua, Otis, and twenty others).[74] The representation of Flather, Jacob Rogers, and Ayer on the boards of local banks repeated the pattern of influence on the local level. Current activity to perpetuate the Boott might be minimal, but the economic position it had helped individuals gain in the past stood undiminished.[75]

At the next level, and with more singular orientation, the Lowell treasurers, along with their counterparts across New England, continued to meet as the Arkwright Club. In 1919 the Associated Textile Industries was revived, and the Boott joined at a maximum liability of $16,000 per year to insure itself and other members against the effects of strikes. Other organizations served the mills as the occasion arose, as in 1918 when the Home Market Club spent $4,000 to defeat a forty-eight-hour bill in the legislature. The Boott contributed $750.[76] The agents of the Lowell mills met monthly as Trustees of the Corporation Hospital, and after discussing hospital matters,

> the mill agents confer in regard to wages. There is a long standing "gentlemen's agreement" between them as to the maximum rates to be paid to certain classes of operatives in lines of work in which the different cotton mills compete. This agents' association also hires a detective who works in all the different mills in an endeavor to suppress stealing.[77]

The use of a detective, or labor spy, was not unusual among the anti-union cotton industry and expressed the level of distrust management felt toward employees.[78] Given the corporations' indignant rejection over the decades of the idea of *labor* organization, their conduct in this regard gave new meaning to the term "gentlemen's agreement."

As occasions arose, the mills formed temporary alliances for joint action. The Boott and the Lawrence confronted their firemen jointly in 1918; the Boott sought to include the Merrimack when work on automobile fabric appeared available, and C. T. Main offered a study for joint use of steam-generating facilities by the mills which lay along the river.[79] The corporations also worked together through the Proprietors of Locks and Canals, in which each corporation held shares and a directorship. In the twenties, for example, they reduced friction head losses (friction between the water and the walls caused the water to drop, diminishing available power) in the power system from three feet to an inch and a half by "sandpapering the canals, or smoothing the walls to preserve available head."[80]

Despite their attentiveness to the operation of the canal system, the Proprietors ignored one consequence of their seven miles of waterways through Lowell. In 1917 a Commission for the Protection of Waterways, "appointed by the Municipal Council to investigate the continued appalling number of drownings in the city and to devise means to minimize, if possible, the awful toll annually losing their lives in the canals and rivers," reported to the Mayor. It found that since 1880, two hundred twenty people had drowned in the canals, and one hundred thirty-six in the Merrimack from banks controlled by the Proprietors. They referred to hundreds who had died before 1880. Despite an average of six fatalities annually in the canals alone, the commission charged that the Proprietors left them unfenced or inadequately so. Such costs stemming from their operations did not appear in the Proprietors' ledger books. Maximizing available power did, and in 1922 they installed a sixth flashboard on the dam for the first time, raising the river upstream to a new high.[81]

The success of the corporations' efforts to dominate the city appeared in the record of labor relations, particularly in the teens. During this period of widespread labor unrest nationally, few strikes occurred in Lowell. Two Lowell mills reacted to the national trend by purchasing insurance for damage from strikes, riots, and civil commotions, not covered by fire insurance, but the Boott did not share that feeling of panic.[82]

While the Boott did not fear labor, neither did they meet their consultants' recommendations to respect or maintain it. Labor, like machinery and buildings, wears out, but there was no depreciation account, no welfare capitalistic pensions, for workers. During an era notable for its speed-ups and stretch-outs (the *American Wool and Cotton Reporter* wrote in 1928 that no mill should have more than 50 percent of the workers it had had ten years before), a period when industrial accidents and their severity rose dramatically, the mill took no financial responsibility for employees beyond the wage they earned when at work.[83] The directors' reluctance to reinvest combined with Flather's economic conservatism to minimize efforts to institute consultants' recommendations. Instead, Flather cited the smallest alterations as significant; looms, for example, were turned at right angles to the walls "to run better," but he did not admit that otherwise they endangered the building. Even in dealings with directors who knew conditions at the mill as well as he did, Flather deflected criticism of the operation rather than speaking frankly.[84] His moralistic attitude, defining success as an effort of will, added a further obstacle to the already difficult relationship between management and employees.

FINANCIAL POSITION

After purchasing the plant of the Boott Cotton Mills for $300,000, the new owners gave it a book value of $600,000. In 1907 the company increased its paper value, to $1,000,000. In 1917 a 25-percent stock dividend raised it again, to $1,250,000, another means by which to increase the apparent size of the investment and to benefit current stockholders at the expense of future investors. These moves diminished the apparent proportion of return on investment through dividends. Both the records of machinery purchased during these years and Flather's comment on heavy borrowing weighing down the Boott indicate that these increases reflected the results of borrowing, not reinvestment.

Robert W. Dunn and Jack Hardy criticized such practices in 1931 in their book, *Labor and Textiles.* They singled out stock dividends, buying bonds out of gross income, high salaries for directors and executives, and corporate income used for loans to investors for their individual benefit as practices which distorted the apparent situation of a company. They identified these as practices which pushed workers to maintain a rate of return based on the artificially increased valuation created without increased investment.[85] The Boott's conduct after its reorganization revealed all these techniques. In addition to the watered stock, the mill loaned its prominent director's Bemis Bag Company $300,000 (at a discount of 7 percent) in 1921, for example, and routinely purchased bonds and other such paper, $975,500 worth in 1922 alone.[86] Clearly they handled their investment in such a way to provide maximum return in the short term, to obscure profitability, and not to develop or maintain either the equipment or the labor force for the long haul. Bemis's role at the Boott offered further opportunities for confusion of interests. He invested heavily there because his Bemis Bag Mills in ten cities bought Boott ducks, drillings, and sheetings at a great rate.

The close involvement of the two companies presented him with opportunities for various manipulations of prices to serve his interests.[87]

According to one analysis based on reported figures, the Boott returned 8 1/2 percent on its inflated valuations throughout its initial operation after reorganization, 1905–14; the percentage of the $300,000 actually invested came to 17 percent. It averaged 40 percent return on investment from 1914 to 1917. Company figures show profits as high as $179,550, or over 50 percent of actual investment, as early as 1907, and in 1915–17, $290,110, $431,898, and $652,607 (twice the investment the directors had brought). None of the difficulties the consultants found at the Boott prevented it from making money, nor did faulty office machines prevent them from recognizing the fact. As presented to the board of directors, the profits of six years amounted as follows:

1915	$342,648.44
1916	$463,241.14
1917	$652,785.34
1918	$565,448.03
1919	$674,183.26
1920	$1,716,741.13[88]

According to other figures from the mill, "profits after all adjustments" reached $777,000 in 1919. Furthermore, the net-quick capital, or ready-money account Flather favored, rose from $290,000 in 1905 to $2,562,000 in 1929. All of this helped explain how Frederick Ayer, a major investor in this and other Lowell mills, upon his death in 1918 had transferred away $13 million from his holdings and still left an estate of $18 million. Even after a decline, felt industrywide, in the twenties, "average [annual] earnings for 25 1/2 years were $253,111.68, or 25 percent on amount invested by stockholders." Despite the impressiveness of that record, it appears likely that without the stock dividends and other machinations, the percentage realized on investment would be several times higher, nearly 85 percent. The Boott Mill during this period performed as an incredible money machine.[89]

While making these profits the Boott could spend up to $100,000 a year on advertising, it could buy out the neighboring Massachusetts Mill and its waterpower, and it could install new turbines and other power equipment. However, it did not make the repairs to plant and equipment recommended by every consultant as essential to *continuing* operation and to the health and well-being of the employees.

The period from 1905 to 1926 was a great time to make money in textiles, and in 1925 Flather was credited with transforming the Boott from a "physical and financial wreck" to a "rich concern" by the *Ameri-*

can Wool and Cotton Reporter.[90] Flather marketed his products so aggressively that the selling house only kept the books, according to Frank Bennett. By the twenties, Boott toweling sold not only in large venues but even in rural grocery stores and drug stores.[91] The Boott investors who had seen continuing prospects for profit in the old mill were proven correct despite all the dire predictions of the experts. At the same time, the problems of overcapacity which had plagued the industry virtually since its inception continued and even increased, a situation many within the manufacturing community noted.[92]

LOWELL AND THE SOUTH

Yet questions of the Boott's future did not turn on events in Lowell, or even decisions in Boston, but on national economic developments. Textiles had remained strangely distanced from major economic trends of the turn-of-the-century period. Trusts and oligopolies, such as those in steel, meat, sugar, railroads, chemicals, and other major industrial sectors, did not protect textile mills from the impacts of competition and market forces.[93] While other industries basked in the warmth of government regulating efforts in the twenties which limited competition and favored big businesses, textiles operated as in the past, with few mergers or acquisitions, unrestrained by law or impulse, maintaining the conduct which had made it a boom and bust business from the start.

The textile industry was exceptionally easy to enter. The potential owner witnessing the profits of good times in the business could easily afford a plant and equipment. For example, when production peaked in 1923, investment growth continued and the number of spindles in place did not peak until 1924, a "most unfortunate disjunction." When duck for automobile tires became a major source of profit in the twenties, tire companies took over the manufacturing of tire duck, leaving textile mills out in the cold. Foreign competition also affected the industry; by 1918 Japan outsold the United States in the Asian market. Each of these developments left the existing companies fighting over the smaller remaining pie. The standard textile response was to press the workers for relief, to make more cloth, and finally to close temporarily or permanently, abandoning the comparatively small investment if times became too tough.[94]

While Lowell had originally benefited from the available power, transportation, capital, soft water, and closeness to the financial center of Boston, by the twentieth century all these factors either had been overcome by southern sites or had ceased to play a determinate role. By 1920, various parties indicated that the South held the advantage with regard to proximity to the cotton, cost of power, tax burden, and labor laws. As time passed, however, each of these elements either proved insignificant or diminished as a proportionate cost. Quality of management in the two areas received conflicting reviews. One crucial point, in all analyses, came to be the cost of labor, clearly higher in the North than the South. However, it was evident in the twenties that the differential did not result from the role of unions. Textile engineers Lockwood-Greene and Company endorsed Lowell in 1920, finding it "a splendid cotton manufacturing centre and in our estimation far more desirable than Lawrence, New Bedford, or Fall River."[95] In fact, "Lowell has for the most part had a reputation for relatively little labor trouble," as Margaret Terrel Parker wrote in 1940. Textile workers were among the least organized employees in the early twentieth century and had very few notable successes before the flight South was well under way. Only 1–2 percent of cotton workers were union members, and in the late teens the United Textile Workers actually had more members in the South than in the North.[96]

While the mushrooming of the industry in the South had created some anxiety after the 1880s, concern had largely abated by 1912, when Copeland, the authority on the business, observed: "These fears have now almost entirely disappeared, since by readjustment and economy the New England manufacturers have given evidence of being able to keep their foothold." Unfortunately for his reputation as a prognosticator, the rush to the South was about to begin. Between 1912 and 1918, the Middlesex (woolens) and Lowell (now Bigelow) Manufacturing (carpets) left town. While in 1923 all the *cotton* manufacturers among the original corporations were still in place, 1926 saw the Hamilton go into receivership, and by 1930 Colonial Textile, occupying the Bigelow plant, quit, Nashua Manufacturing bought the Suffolk, Merrimack purchased the Tremont and destroyed most of the plant, Pepperell bought and discarded the Massachusetts, the Appleton moved to South Carolina, Ipswich Mills, new owner of the Middlesex, sold out, and the once mighty Lowell Machine Shop was razed. Charles T. Main noted in 1926 that the industry was hampered by a general depression, high prices for cotton, decreases in export trade, and increases in imports of cloth, providing proximate causes for the shifts. Of the giants, only the Merrimack, Lawrence, and Boott remained, and the former also ran a southern operation (as had the Massachusetts, indicating that such a course offered no guarantee of success).[97] In Lowell, product lines lost to the South included cotton blankets, flannels, ginghams, chambrays, shirtings, and sheetings.[98] By one count, 855,000 spindles left Lowell alone by 1929 and $100 million of New England mill capital went South between 1923 and 1927. As a result, again according to C. T. Main, southern mills which had been built largely with northern money had paid their debts between 1914 and 1921 and only by 1922 emerged as competitive. Ideal conditions had led to relative expansion, North and South, between 1900 and 1924, of 15.1 to 20 million spindles, or 32 percent, in the former, and 4.4 to 16.3 million spindles, or 270 percent, in the latter! Southern conditions led to 60–100 percent more operation there, with its mills running steadily for 110 hours a week on base production and the North running for periods of peak demand. The industry had found a way to persist along its original course.[99]

The failure of agriculture in much of the South in this period offered a new source of labor. The difference lay not in unionization, but in costs. As Dunn and Hardy put it, in the South the industry found "cheap and contented labor." As Main explained it, southern workers came to the mills from a background of "privation" and appreciated the work; they accepted it on the basis of a "family wage" so low it would not maintain a standard of living acceptable in the North.[100] Despite the impoverishment of Lowell workers, despite the necessity there for family work, southern laborers could be had still cheaper and stretched still further. English-speaking, they would be easier to train than the Southern Europeans who were available and who offered the alternate inexperienced (inexpensive) workforce in the North. British observer T. M. Young claimed that the native-born southerners, unlike the northern workers who were 60 percent foreign-born or of foreign parentage, offered ideal qualities for employees:

> The practically universal use of the English language in the South is a great protection against labor disturbances and thus far labor agitators and other trouble-makers have gained but little foothold in that section. This is a great advantage to the owners because they have much less difficulty in introducing new methods and find the employees generally more tractable when they are free from outside domination. Much of the labor recruited locally by the Southern mills is drawn from the surrounding country and has never been employed on other than farm work.

In Young's view, management domination of the workers was more important than their skills, interference with the workers was foreign, and ignorance of the workers was necessary to total control. In the South, the industry sought to escape the attitudes that it had itself produced among experienced workers. The workers' background of "privation" was used in combination with the mill owners' political power, their mill-village control, exclusion of "outsiders" (experienced workers), and the dependency associated with a "family wage" to regain the control that had been the essence of Lowell-style development.[101]

The Boott had a variety of connections to southern textiles. Director Bemis had mills in a number of cities in the South. The Ayers's heavy involvement in the Merrimack connected them to its Alabama mill, as it did Flather, who was managing both the Boott and the Merrimack. Calculations of advantage regarding mill location pervaded their discussions. A letter to the stockholders of the Tremont and Suffolk Mills in 1924 claimed that southern mills were making napped goods in competition with them at one-half the labor cost. In the same year a Boott memo found southern wages represented a 125-percent differential, a difference of more than two to one in comparison to Boott wages. A more measured analysis to the situation appeared in the Main study, which found a 13-percent advantage in labor costs for southern mills.[102]

Ever since its reorganization in 1905, the Boott had maintained reserve accounts for various purposes related to holding undispersed income. One, a Depreciation Account, had been raised by $120,689 in 1914 in order to reach the arbitrary level of 5-percent annual depreciation of the valuation. For the year 1917, 10 percent was added to the account due to the effects of night work. In part because of the large amounts of cash reserved, in 1922 and 1923 the Boott searched for a southern mill to purchase for operating duck looms it felt it couldn't run profitably in Lowell. In 1925, the directors were again "considering a southern mill."[103] The effect of the attitude these searches implied became clear in a statement made by the treasurer to the directors in 1927: "The custom of paying dividends out of past earnings for the last five years, which when taking depreciation into account were not earned, it is believed will be continued and increased when warranted by economic conditions." Despite the knowledge that to pay dividends at the expense of depreciation adversely affected the chances of the plant's survival, the procedure would be continued.[104]

In 1926 an auditor had told the Boott that from an accounting standpoint, the Depreciation Reserve accumulated exceeded the book value of the plant. After that time, the new policy was to continue the reserve of $2,711,767.80 but not to increase it. If the company were to spend on the plant each year the amount which previously would have gone into the reserve, explained Flather, it "would continue the expectance of life of the plant about as it is now"—which was to say in operable condition. However, twice in the next two years the treasurer explained the nature and significance of the new policy:

> Not adding to the Reserve for Depreciation is satisfactory from an accounting and book values standpoint, but from the standpoint of prolonging the life of the plant year by year, an amount equivalent to a fair depreciation charge should be either reserved or spent on the plant annually. If each year such a sum should be expended on the plant at Lowell, it would be

> only where such expenditure were of special usefulness. Any large expenditure for equipment of a general nature would better be expended in the South.

More simply, regular expenditure on the plant at Lowell "would not pay," while the same money "spent annually in the South would pay better."[105] At this point the die was cast: the Boott Mills could not endure in Lowell. The profit the mill and its employees had supplied the shareholders and accumulated, in part in the Depreciation Reserve, would not be used to perpetuate this operation.

In 1929 the Flathers' desire for southern investment of Boott profits appeared as a request to the directors to approve buying or operating the Stirn corduroy plant in the South, with finishing to be continued at its factory in Williamstown, Massachusetts.

CONCLUSION

After the reorganization in 1905, the Boott Mills's directors brought in consultants for repeated analyses of their operation. The need for them reflected the new style of mill management: Cumnock had been the consultant for others, not the employer of them. However, consistently ignoring the advice of those consultants, particularly with regard to working conditions and employee treatment, the directors revealed their intention to continue the Lowell tradition of limited investment, tight cost control, and steadily increasing workloads. In fact, at this point the enterprise relied primarily on borrowed money on the one hand and increasing exploitation of labor on the other.

The mill's condition presented an obstacle to accomplishment which affected all involved. The attitudes of management toward reinvestment, and as a result labor toward the corporation's purposes, were notably poor, and so were the earnings of the workers. Management left the workers and their interests out of planning; workers who came, or stayed, responded to their treatment with little interest in anything more than minimal fulfillment of assignments.[106]

In such a context, the purchase of a few new machines from time to time had little impact. If the floors were so deep with tobacco-spittings that loom-fixers were reluctant to do their work, it made little difference whether or not the looms involved were current models: they were likely to be neglected. The train of difficulty caused by poor operating conditions extended in all directions, even up the chain of command. If looms were not fixed promptly and adequately, scheduling became impossible, production lagged, seconds increased, and everyone was frustrated.

Management pursued several answers to these problems. The ideal of increased production through study and analysis proved repeatedly tempting. Its vogue coincided with the shifting of management's background from people with processing expertise to those with accounting and marketing skills. The extent to which the consultants produced a continuous complaint, rather than revealing changing conditions, indicated that they were not often heeded, that their advice was sought almost as a talisman, something which might cause circumstances to change, or conditions to improve, without the expenditure of meaningful amounts of money.

When new technologies became available, management did not need consultants to bring them to their attention. The managers identified the Barber devices, automatic looms, and Casablanca's long-draft spinning independently as significant inventions. Despite Flather's background in innovation at Pettee and the LMS, he brought little new equipment to the Boott.

The major impact of the Flathers came from the level of their involvement. Under them the mill continued to produce the diversity of materials warned against, but it operated under close, if not expert, supervision. The Flathers actively pursued new markets rather than leaving them to be divided up by a selling house representing several factories. They identified niches and produced materials for them. They pursued new opportunities, such as those associated with the mushrooming automobile industry. While other Lowell mills also turned to automotive fabrics (the Merrimack survived on these for decades), the Boott maintained a diversity of products in addition to the big sellers of a given year.

With the Flathers, the Boott owners had achieved their goal of hiring managers dedicated to the interests of the company. This family dedicated itself to the mill's success as to a cause. They developed materials which they could produce profitably, particularly in wartime, they sought out new markets, and they emphasized advertising and salesmanship, all while pouring energy into an operation of which they were, in small part, owners, and for which they had taken, in a major way, responsibility. Flather made it a moral issue. For him, private interest and corporate financial success were inextricably joined. Private interest and public good, the claim of Henry Miles in the early days, again appeared to be one, and the identity came in the form of Boott Mill success.

Part Four

Depression, Continuity, and Change, 1931–1946

Chapter Nine

Factors Affecting Survival

THE DIRECTORS had placed the mill in a curiously fatalistic position, one in which success could only be temporary. Despite the nature of the situation, Flather continued to deepen his financial involvement in the Boott. His personalized view of its success or failure led him to increase his participation as an investor, though he knew its days were numbered. The conflict between the board's decision not to perpetuate the operation and his internalization of its fortunes presented a new occasion for the tensions that plagued him. Like a character in a Greek tragedy, he persisted despite the preordained conclusion, the certainty, in his terms, of failure.

The extent of the Flathers' involvement in the Boott, and their attitude toward the mill, could be seen in 1939 when the senior Flather called the board's attention to a facetious, but meaningful, editorial in the *American Wool and Cotton Reporter.* After being refused its request for names and addresses of the mill's overseers, it described the Flathers:

> He is at the mill every morning and at his treasurer's office every afternoon, and in constant touch with the selling agents all of the time. We are told that this mill has no overseers, no second hands and no superintendents—that the old-time operatives with steady work and with good pay, need no overseers, and act as their own second hands. Two of the treasurer's own family are very active in the business, and are at the mill first in the morning and away from the mill last in the evening. They are all the overseers, and all the second hands, and all the superintendents.[1]

While I cannot be sure what prompted Flather to read this into the Board of Directors' Minutes, speculation is irresistible. Indirect bragging, defense against implications of nepotism, and resistance to continual consultants' reports that the mill had too many employees per spindle all suggest themselves. In the context of his formal relationship with the directors, as reflected in the minutes of their meetings, there is a strain apparent here, something out of character. The times had been trying, his appointment of his sons could make him vulnerable, his habitual aloofness made explanation difficult. Irony also appears in the episode: the Flathers did not supply the operational expertise for the mill; supervisory personnel did. The Boott relied heavily on employee knowledge and skill.

TACTICS

In the early thirties the Boott operated as a variety mill, producing "towels, moleskins [used in sheepskin-lined coats], duck, seamless grain bags, Bedford corduroy, osnaburgs, damask, twills and drills, corduroy, velveteen, scrim and marquisette." These goods, made with an average yarn count of #15 1/2, qualified the Boott as a coarse mill once again and placed it in direct competition with southern mills dominating this range of goods.[2] Conditions in the mill at this time would not support the production of finer materials.[3]

Scrim (for curtains), corduroy, and toweling predominated, with production of some 11 million yards apiece in the later thirties. Rather than sending these products to finishers, management, particularly Frederick, experimented with bleaching and dyeing the cloth in-house. The Boott was not the first producer of corduroy in the country, as it claimed, but by 1935–36 it could call itself the "largest individual producer of corduroy in the entire United States." While it sold toweling to many customers, however, it sold corduroy almost entirely to one, leaving the mill open to disaster in 1936: "Then we had a flood, and our customer got terribly scared." Whereas they had run 1,600 of their 2,000 looms for this customer, in 1937 he affiliated with southern manufacturers.[4] This event indicated the precariousness of their situation in several regards. Producing for just one customer, offering nothing he couldn't obtain elsewhere, and unable to significantly undercut the competition on price, they became vulnerable. Although they operated as an order mill, the scale of the Boott's operation denied them the prompt flexibility of Scranton's Philadelphia manufacturers. Toweling and scrim sold under the Boott name to many customers after finishing and sewing overcame some of these problems. Although Frederick rewrote history as early as 1937, claiming that, "if we had not lost our staple runs because of high pay," in reality the array of obstacles arising from the absence of continuing reinvestment created a situation which the flood aggravated despite low wages.[5]

Circumstances (the loss of the corduroy customer) combined with intent and events to greatly reduce the variety of the mill's production. In the late thirties products included damask (defined as cotton table cloth with stripes); scrim; and toweling, for drying dishes and uses where paper towels later came to serve. An exclusive franchise for an absorbent agent, Sorbtex, was acquired to aid their performance and marketing.

The Boott distributed towels and toweling "to wholesale, retail, and chain stores, hospitals, hotels, institutions and laundry supply houses" nationally. A cutting and sewing operation finished towels and curtains manufactured "for durability." They also finished their "Tiffany grade" corduroy produced as part of a group which had dominated that fabric, "the old big 4 manufacturers (Boott, Crompton, New York Mills and Merrimack)." Making goods which the mill could both market on its own and sell on the basis of name-recognition fulfilled long-time advice. Both the toweling and corduroy represented materials for which they anticipated steady demand for an unchanging product. Only the entry of "new-comers" (by which Flather meant mills in the South) into these fields altered the situation.[6]

The Boott increased its efforts to finish its products as one way to improve its competitive position. Management gradually abandoned grey (unfinished) goods as offering little opportunity for profit. By 1940 it dyed, bleached, and mercerized scrim and also bleached and finished towels, including calendering linen-finish towels (7 1/2 percent linen). Performing its own finishing separated the Boott from southern operations, most of which sent grey goods north. While this processing advanced the mill's cause, it also gave rise to reservations. As the Treasurer's Report noted in 1940, it proved "quite trying to some inelastic features of the physical plant and of the personnel." Furthermore, when Frederick discussed the processes in a directors' meeting, it appeared that Director Roberts knew more about calendering and alternate techniques than Frederick, management's expert in the field.[7] Limitations of plant and personnel could indeed interfere with the success of this particularly high-skill area.

Finished toweling represented 5 million yards of production in 1937, and by 1940 it led *weekly* production at 208,000 yards, followed by 80,000 yards of scrim and 3,500 yards of corduroy. A new product, however, combed drill uniform cloth for the Navy, was at the same time changing the balance of what was produced at the Boott. Production of the Navy cloth had grown from 2,000 yards in 1938 to 458,000 in 1939 and to 2,780,000 yards in 1940. By 1942 it reached nearly 14.5 million yards per year, or 273,000 per week.[8] During the next two years the pattern continued as production rose to 500,000 yards per week in 1943 and peaked with 48 million yards delivered in 1944.[9] One of every two Navy sailors wore Boott twill. At a price of cost plus 5 percent, the Boott suddenly found itself in a new situation regarding steady and predictable profitability on a large-scale run.

War production enabled the Boott to reduce 100 yarn numbers to

15, 200 styles of cloth to 11 (3 towelings [1 for the Army]; 2 corduroys; 2 scrims; 2 ducks [1 Army, 1 Marine Corps]; 2 drills [both Navy]), greatly simplifying the task of keeping production "in balance." Demands on planning and scheduling personnel diminished. Wartime made it hard for the Boott to maintain a workforce, however: "We are not so well covered in the lower strata of skilled and semi-skilled productive personnel, but that is another story, and is being provided for by turning to training women."[10] Wartime rekindled the old Lowell practice of training women for increasing numbers of textile jobs, now as replacements for workers lost to the draft or to better wartime employment.

As the war continued, the Navy demanded colors other than white (in which faults became more visible), stiffened inspection, and cited problems reflecting on every department and causing intensified efforts to improve quality. At best, though, the cloth won the Navy's praise, as well as that of Williamson and Dickie, makers of work clothes, who found it superior to army suitings available elsewhere.[11]

All in all, the Boott not only did very well during the war but was left in a potentially strong position after it. They had not spent the war years making specialty fabrics with little civilian market, but they had simplified their production and had specialized on drills and toweling, for which a strong demand persisted. Despite an excess of manufacturing capacity nationally, a predictable result of the profits available in wartime, the Boott could withstand the resulting cloth-dumping, for example by Cannon in 1946.

THE MILLS

The Boott faced the accumulated effects of years of its neglect of its plant. Buildings once admired as part of the showpiece that was early Lowell had, along with their successors, become increasingly light and flimsy in comparison to the increasing structural needs of a modern cotton mill. At a 1945 tax appeal, experts called the buildings "poor," with walls generally out of plumb, floors out of level lengthwise and crosswise, windows in bad condition, toilets small and scarce, rooms dark, narrow, columned, and low, vibration excessive, sprinkler heads the old-fashioned pendant type, no passenger elevators, inadequate coal storage, undrinkable water, and moisture from penstocks and runoff from the yard rotting timbers and floors, and humidity rotting frames and sashes. When the tax commissioner asked incredulously,

"You have never done anything about changing the conditions of the building in all these years?," Frederick replied disingenuously that they could only have built new buildings. Management instead removed towers, installed tie-rods to reduce cracking in walls and swaying of towers, and ceased flying the flag "to reduce sway." When Director Frederick Ayer had asked facetiously in 1934, "Don't you think that is extravagant, spending 20 dollars on the plant?," Flather assured him that money would be spent, but it was not. Only dribs and drabs of funds were allocated to patch and mend, none to rehabilitate or perpetuate.[12]

The consequences on the operation were many. A foreman, a master mechanic, and a staff of seventy-two men had to be kept to maintain and repair the complex, thirty for machinery and forty-two for the buildings. They applied a "gondola [railroad] car" of steel plates to checked or split beams. A ten- to twelve-man crew worked on the roofs in all dry weather, summer and winter. Two men worked full-time attempting to fix the windows. The Sisyphean tasks of these crews offers a sad metaphor for Boott mill work during this period.[13]

In addition to repair needs, the buildings harmfully affected all aspects of the operation: "Dust control is very important in the interests of quality and in the attraction of personnel. We are very limited in our building to use that because of the multiplicity of ducts which would be required in so many rooms which are so low-studded it would interfere with our drives, belting, shafting and humidity." Employees also suffered because the same factors limited the mill's ability to improve conditions with regard to heating, cooling, lighting, or noise. Some machines actually became wet during damp weather, and makeshift efforts to solve the problem by ducting steam from wet areas to others brought unsatisfactory results: "The employees objected because whereas in ordinary cotton mill procedure the humidity is cold, this was hot, and they objected to a hot blast coming out from the ventilating duct, and we had to remove it." Supervision of the plant also suffered from the buildings' impediments: "Our departmental supervision is seriously interfered with. If we had a modern mill, we would have three major manufacturing rooms, a carding room, a spinning room, and a weaving room . . . and you would have three foremen or supervisors or overseers . . . whereas we have some fifty overseers because we have to have one in every room." On the other hand, where rearranging spinning frames transversely to the buildings would have improved supervision, columns of posts prevented it. In 1945, Flather deemed the property "not worth 5 cents."[14]

The Flathers continued to encounter obstacles only new investment could overcome. They noted that "to increase work assignments [for spinners] would require longer frames." John Rogers cited further limiting factors associated with the buildings:

> Because of the layout, the floors and the difficulty with posts and columns, we had 3 loom fixers . . . averaging 63 looms apiece. A southern mill without any building restrictions and a one story weave shed, should run . . . 90 to 100 looms. . . . We have loom fixers working on two floors . . ., and obviously they are going to refuse to handle as many looms when they have to go up and down stairs every few minutes.[15]

The managers were well aware of the difficulties inherent in their plant. Floors were "oil-soaked . . . splinter badly, . . . [and] are weak because there have been holes cut through most of them over a period of a great many years in the old shafting days. They are too light for running modern high speed machinery." Uneven, full of travelers in spinning rooms, the floors prevented use of new floor-scraping machines because they leaked water through to floors below. Auto-cleaners could not be installed over spinning frames because of the many obstacles; bearings in spinning and roving frames burned out and started fires because of the uneven flooring. Modern "streamline [sic] production" could not be attempted in available spaces, and 10–12 percent of the employees were involved in transportation rather than production. Poor foundations led to sagging floors and distorted the boilers. Looms had to be rearranged to avoid increasing the lean toward the canal in one building, toward the river in another. Precise settings of machines had become impossible, and conditions dictated that they "did not put in the fastest machines that were available in the thirties." Running Draper Model "E" looms at 150 ppm, they attempted to introduce just sixty-seven high-speed Model "X" looms running at 200–224 ppm in #5-3W: "They shook down the wall of that building at the corner of the river and the wasteway so that it had to be relaid. . . . We would like very much to have more of the Model X looms, but our experience was so disastrous on the building that we have hesitated to go ahead." While other machines produced less dramatic results, most had to run at less than, often a fraction of, standard speeds.[16]

The plant, perhaps never a good one, certainly insufficient and outmoded since the fourth quarter of the nineteenth century, had been allowed to become the Boott's primary limiting factor, blocking modern machinery, supervision, transportation, and workloads, while sapping all efforts and draining budgets for maintenance and repair. Vibrations "became particularly pronounced in the decade of the thirties and in the war period," when three-shift operation combined with heavier machinery and less-skilled help to take its toll on the antiquated structures. One building had to be built "for the express purpose of bracing against the wall of No. 4 mill where the bulge and the lean were." But with regard to an equally long-standing problem, the water damage caused to buildings #1–#4 by the bad drainage in the yard, the tax commission lawyer pointed out, "You haven't taken any steps whatever to relieve that situation." If they could avoid treating a problem on more than a day-to-day basis, they did. As one cost estimator at the hearings put it, "The maintenance of these buildings has been neglected, and the condition of the buildings has consequently suffered." According to consultant John A. Stevens, "These buildings could not be used to any advantage for the present high speed, heavy, improved textile machinery." The absence of planning for joining the mills by the original designers hampered transportation in the operation of the complex as an integrated unit: truckers encountered slopes at every intersection of the many buildings.[17]

THE MACHINES

Discussion of technology after 1931 often turns on questions of conditions: those of the machinery, the processes, and the times. The managers confronted changing circumstances amid deteriorating buildings and equipment. George Sweet Gibb emphasized this aspect of the situation with regard to the implications of long-draft roving and spinning: "Coming at a time when low-cost operation was essential for survival and when labor rates in the textile industry were being boosted by political action, these developments could not be ignored." The Saco-Lowell Machine Shop editorialized: "We firmly believe that the principles of the new deal having to do with wages and working conditions have practically prohibited the further use of much obsolete machinery and have made the question of mechanical efficiency one of the most important confronting mill executives at this time."[18] Both Gibb and Saco-Lowell referred to the federal legislation giving labor new opportunities in the areas of organizing and negotiating. Gibb's hindsight and the builder's contemporary promotional view signaled a recognition of changed conditions, a perception that machine replacement had become even more important than before. The economic downturn presented only the immediate reason to avoid investment in new equipment.

Insurance appraisals show that by 1939, carding machines dated from 1887–1909, with two-thirds from the period 1906–7. Two thirds of the drawing frames dated from before 1905, the rest from slightly later. Roving and slubbing machines were of comparable ages. Slashers dated from 1885–97. Looms were about evenly divided between pre-

1905 and 1905–12. Operations seriously interested in continuing to compete effectively in the industry, and with the South, would not, and were not, utilizing this sort of antiquated machinery. Twenty-year-old looms were seriously worn in the best of circumstances, and no one ever suggested that machines at the Boott were running in such circumstances. Thirty-year-old looms had no place in a competitive operation. And yet this complement of machinery, with minor exceptions, operated until the mill's closing.[19]

The Flathers knew the condition of the machinery they oversaw. In 1932, Flather traveled to Alabama to evaluate the condition of the Merrimack's Huntsville operation. He admired the plant, criticized the personnel for its lack of "technical leadership," and observed that "the looms are dirty to a most unfortunate degree. Their condition must result in much soiled and oily goods." A report on the shape of the Merrimack's equipment in Lowell in 1935 presented a view that would also have applied to the Boott: "A major portion of the machinery in your plant is very old, much of it having been in continuous operation from thirty to thirty-five years; it is naturally obsolete, inefficient, and uneconomical to operate as compared to new machinery." Flather recognized problems, but he couldn't cure them.[20]

In 1935, Frederick told the directors that money needed to be budgeted for machinery, "on account of the competition we are going to be faced with substantial expenditures—for high speed looms." When Frederick Ayer asked, "Does that mean that all of our looms are obsolete?" Frederick replied, "All but one [an experimental "X" loom]." In 1945, Frederick stated that everything had to be replaced as soon as possible. When asked, on another occasion, why they had not "more fully modernized the machinery," he answered, "We haven't had the money." The directors did not welcome requests to spend money on new machines; they complained when repair expenses ran as high as $1 per spindle per year, certainly a minimal level.[21]

During the thirties, they converted some roving and spinning frames to long-draft: one-third to one-half of the frames received the attachments. Since they did not replace the frames themselves, the size of the packages they created (and thus the time between doffs) could not be increased, as they were on new machines, and thus the effect of the alterations was limited.[22] Between 1939 and 1950, the Boott devoted less than $100,000 to new machinery for the mill, and a quarter of that sum went to office equipment (see table 8).[23] They spent just $2,000 to $3,000 per year per department, on average.

Such minimal efforts did little to keep the mill apace of machinery developments; rather, they offered stop-gap solutions to continuing problems. As Frederick said in 1937, "Most of our new equipment is a

TABLE 8.
Equipment Purchased since January 1, 1939, in Dollars

	Office	Weaving	Spinning	Laboratory instruments, Tools
1939	1,259.50	1,887.48	14,035.15	265.58
1940	827.63	...	168.40	944.28
1941	995.15	11,000.00	1,377.75	294.15
1942	1,400.46	...	951.25	2,019.10
1943	990.25	...	...	239.42
1944	26,410.46	...	122.50	376.80
1945	1,441.10	2,293.95	1,525.65	948.50
1946	1,802.80	...	1,075.00	373.84
1947	454.84	...	57.65	...
1948	701.97	...	558.68	...
1949	2,296.29	17,530.82	1,292.81	92.99
1950[a]	...	118.50	...	...
Total	38,580.45	32,830.75	21,164.84	5,554.66

Source: Directors' Records, March 1950, box 37, Flather Collection.
[a]Until March 1.

small percentage of the whole. It is a demonstration equipment."[24] The Draper "X" loom offered an example of this behavior. It was a faster running loom than the "E" and "K" models. Boott Mills bought one in 1933, and by 1940 they owned 78, a drop in the bucket of their 2,000+ looms. They also tried to run existing looms at higher speeds, but little seems to have come of that effort, and no wonder, considering the worn-out machines on which it was attempted.[25] Perhaps most important, while aware that their looms were slow, inefficient, and often of the wrong size for the goods, the Flathers were philosophically opposed to remedying the situation. After buying a new Draper terry towel loom, Frederick observed: "The price was so high that we couldn't get our money back in twelve years. We ought to get our money back in a year or two," an absurd position. When the mill closed in 1954, it still ran belt-driven looms as a result of such arcane thinking.[26] When, in 1954 and 1955, the equipment was appraised and sold, practically everything went for scrap. Twenty-six dollars a ton defined its quality and value.[27]

WORKING CONDITIONS

The Ralph E. Loper Company, which had studied the mill previously, returned in 1944 and found that the mill's machinery and conditions still worked against effective operation and stymied production. It found Boott labor costs per pound of product higher than the South's "*after making allowance for the difference in wage rate*" [emphasis original]:

> Your excessive labor cost which is both the proof and the result of operating conditions being far from satisfactory. . . .
>
> 1. The work runs so poorly that it is impossible to maintain a normal rate of machine production or to assign normal work loads in all cases. . . .
>
> 2. Employees lack sufficient initiative to perform to the best of their ability, in certain instances, under present operating conditions.

Loper blamed management for the former, labor for the latter, although they had not produced the conditions. "Operating conditions" represented the results of the mill's history.[28]

Machinery was not only old, it appeared to be driven at practically random speeds. Looms running at 137 picks per minute (ppm) rather than the expected 160 ppm prevented weavers from earning the "standard wage" and cost the mill production. Working conditions "considerably below normal" made employees first "indifferent," then "antagonistic." Ultimately, "the basic responsibility for low employee productivity in Boott Mill rests with management because it has allowed conditions to exist which are detrimental to efficient operation." Loper charged that the operation suffered from the lack of a plant manager expert in cotton manufacture and recommended that someone knowledgeable, established, and able to work in harmony with labor be found and added to the staff. He also noted a need for a better stock, straight cotton (without waste), in order to improve the breakage situation that was hampering production and frustrating labor. End breakage on spinning in combed yarns at times doubled acceptable frequency. The procedure of mixing in waste from processing offered false savings, an indication of the ways in which managers not expert in the field could appear to benefit the mill through frugality while actually creating production problems and discouraging employees. Dirty and nonuniform spinning frames caused excessive broken ends and also produced inconsistent yarn. Slow-running frames and looms frustrated workers through unrealistic piece-rates and revealed poor maintenance, improper settings, and incorrect parts. Loper concluded by suggesting "that management would be in a very vulnerable position in case outside parties are called in to settle disputes."[29]

One technological change affecting conditions did not stem from the choices made in Lowell. Spinner Emma Belisle complained in 1950 that the cotton was not as clean as before, left more dust on the frames, and was more difficult to run. While management suggested that workers in preceding processes could cause bad work, it also noted (privately) that, "the machine-picked cotton now reaching us has more dust and dirt in it than the hand-picked cotton and it is not surprising if more dust is now occasionally seen. However, our management knows the problem and has taken steps to make corrections where and when necessary."[30] If this last were true, of course, the complaint would not have been made.

Training and Skill

Another study done in 1945 carried unsettling implications about the nature of work and relationships among workers at the Boott. Referring to "accepted and well understood" procedures and "well established" practices, it outlined expected conduct: Workers on different shifts should show mutual respect and turn over operations in good order. Machinery should run continually; smoking was prohibited; "It is well established and understood that supervisors are responsible for discipline and have the authority it implies."[31] If we operate on the assumption that rules are not made to bar things which people do not do, we can assume that basic understandings were not "well established."

By the late 1930s the Boott's chronically high labor turnover increased as workers moved to higher-paying jobs at other employers such as U. S. Cartridge. Not surprisingly, attrition was greatest among the more desirable employees, those most able to fill positions elsewhere. Of those that remained, Flather noted, "Some of them we once would have allowed to go with our blessing, but now we want them to stay even though they are not of high efficiency."[32] Early in the war, turnover represented 3 percent of the workforce per month. Decreased rates accompanied government moves to reduce labor mobility.

Replacing workers not only required extensive training programs but also made the mill even more susceptible to dislocations within its production system. In 1943 John Rogers described the necessity of laying off loom-fixers, not for their performance, but because turnover and absenteeism in carding created a bottleneck.[33] Recognizing the difficulties faced, management in 1941 had called on the Office of Production Management, Northern New England District, Training Within Industry, Branch of Labor Division (OPM, NNED, TWI, BLD) to assist with the mill's effort to confront the situation. Traditional approaches of assigning new help to experienced workers failed to meet the demand, so "schools" for loom-fixers and weavers refurbished and ran previously idle looms. This system worked well but became insufficient partly because of the amount of supervisory help it required, partly because it too did not work quickly enough to meet the needs of the mill, and partly because the operation lacked standard machine settings and practices.

The advent of time and motion studies at the Boott during the forties led to careful measuring of the elements in each job in the mill. It combined with the training program to produce careful analysis and descriptions of them. Management hired a New York engineering firm, Albert Raimond and Associates, to measure and define each operative's tasks.

Overall, the effort aimed at increased definition, standardization, and specialization. Positions treated included weaver, loom-fixer, battery hand, filling boy, and loom cleaner. Efforts focused on weaving and loom-fixing as areas in which certain tasks could be specified and taught in a standardized way. The importance of uniformity of approach was emphasized, particularly across shifts.

As part of this analysis of training needs, several "Exhibits" described the particular tasks supervisors had brought up as both measurable and crucial. For weavers, they outlined the skills identified as essential:

1. Tying the weaver's knot
2. Drawing in warp ends
   a. Drawing in new ends
   b. Repairing a broken warp thread
3. Finding the pick
4. Starting loom properly
   a. Properly placing shuttle in shuttle box
   b. Letting back warp yarn
5. Keeping correct tension on cloth and warp
6. Inspecting cloth for defects
7. Inspecting warps to prevent stoppages.[34]

They then prepared a sequential approach to those operations which would lend itself to teaching.[35] Since managers, then, and scholars, since, have often referred to the "unskilled" nature of much textile work, including weaving, it is worthwhile for us to consider the description of "Key Points" in the performance of one aspect of the several tasks associated with a weaver's work. The OPM, NNED, TWI, BLD described the procedure of drawing in a broken warp end. The task required the weaver to move the beater to the proper position, use the stop motion to find the broken end, repair it with weaver's knots, draw it through the correct drop wire, heddle, and reed dent with a weaver's hook, reposition the beater, and restart the loom. Remember the situation of the weaver. One of a number of looms he or she tended had stopped, signalling a problem. In the midst of the indescribable cacophony of the weave room, the weaver had to note the idle machine, hurry to it, and determine the cause. Piecework wages halted as long as the loom was still, and the fear that another loom would need attention while the weaver dealt with this problem carried the threat of a multiplying economic impact. Not only dexterity and good eyesight were required, but ability to work under pressure. Error in any of the steps described would create further difficulties. While the job might still not be seen to require brilliance, it becomes clear that it did involve skills, and that the amount of pressure on the weaver's performance increased rapidly in proportion to the number of looms tended.

After finding and repairing a weft break, which required greater skill, the weaver had to turn the loom backward to the harness position at which the break occurred ("finding the pick"). In order to avoid a fault, the cloth on the take-up beam had to be "let back" to the point where a resumption of operation would produce cloth with neither a thick nor a thin spot. This was done not by rule or gauge, but by feel and experience. More complicated looms and cloth accentuated this problem, and the Boott's 20 harness dobbies ranked high in the level of difficulty presented. Similarly, adjusting the warp-tension mechanism had to be done with judgment as the size of the warp beam decreased during weaving. These and other aspects of the assignments made demands which were completely met only by well-trained and experienced weavers.

The concepts of "maximum production" and "highest quality" defined the two conflicting pressures on labor. Several points in the job descriptions revealed the requirement of judgment and expertise generally denied this classification of workers. In loom-fixing, the fixer confronted the particular difficulty that alteration of any one of the loom's seven motions or its settings had an effect on all the others. For example, simply adjusting the binder on the shuttle box which decelerated the shuttle after it passed through the shed changed the timing of its arrival at the picker and the amount of power required to return it to the other side of the loom; it also affected the timing of the shedding and beating motions. Many of the adjustments required careful judgment, and even those which were supposed to be checked with a gauge required increasing skill as looms aged and wore. In the case of the

Boott, this included all production looms. Work also became more difficult when faster machines were introduced.

The need to prepare the training programs indicated the strength of the informal ways in which workers had learned the jobs previously. Family connections, as well as others, such as neighborhood and nationality, had enabled a prospective worker to come to the factory with an educator already lined up. The demands of the war on the country's productive capacity eliminated the luxury of time which individualized instruction demanded. The Boott's inability to compete with other employers for the available talent required it to come up with new methods to train replacements for the employees continually lost to better jobs. The necessity for careful efforts to improve training, as well as the descriptions it produced of tasks, reveal a level of complexity for textile work which belies the standard accounts that dismiss it as unskilled. Finally, the training problems revealed the inability of "scientific management" to produce a level of simplification and/or specialization making the work easily graspable.

Suddenly the assumption of low-skilled work appears less safe, the descriptions of tasks found in job descriptions less simple, perhaps less complete in their ability to convey the requirements to an inexperienced reader. One is reminded of the phrase from the instruction manuals regarding judgments "not reducible to numbers" associated with numerous textile chores.[36]

The Flathers offered their own testimony regarding necessary skills, as well. In 1937, advocating more super-draft roving machinery, Flather cited its advantageousness in the face of "a shortage of skilled help which is quite serious." Inability to attract the skilled during the Depression boded ill for the future, and the recognition of the need underlined its importance. When the mill introduced finishing operations, he evaluated the work of the corduroy cutters: "In the process of cutting, the operative must acquire requisite skill with the delicate guide, knife, and stop motion, but this skill is sure to materialize faster than in learning to walk; but not quite as rapidly as one learns to steer an automobile." Unwilling to hire trained cutters, presumably because of their knowledge of their significance and concomitant salary demands, the Boott had to put up with the training period. When wartime industries drew away workers, he spoke of robbing one department of "skilled weavers" to aid another, noted that "good labor is scarce, especially experienced spinners and weavers," and "we certainly have increased our appreciation" of experienced workers. Inexperienced help even at the level of trucking was said to have a bad effect on operations, and in weaving, "looms have taken a beating from the unskilled changeable employees of war-time." Full expression of the conundrum of constant improvement in "labor-saving" machinery without elimination of skill came in the testimony of Frederick:

> The modernizing of machinery usually carries with it a higher degree of automaticity so there is more to be done by the machine and less done by the individual. In other words, speaking of it in terms of time study or load factor, the individual employee works fewer minutes of the hour and with less personal physical effort on the modern machinery than on the older machinery.
>
> Q. Is there greater skill required?
>
> A. The modern machinery does require greater skill, yes. It is a more delicate machine and a more automatic machine.
>
> Mr. Ready: That is one advantage you have anyway, isn't it?

The Boott's "advantage" was its lack of new machinery and its skill requirements. At bottom, skill played a steady and at times increasing role, especially at this mill.[37]

The Flathers recognized the importance of their workforce's skills. During World War II, the Boott trained 5,000 "green hands" to produce 1,200 useful employees. The workers' efforts remained pervasive in the operation: "By participating in the recruitment and training of new workers, coaxing the aging machines to run, and coordinating the process of work, mill workers contributed to managing the mills in ways crucial to production but taken for granted or ignored by the mill managers."[38] Two recent comments by Lowell mill-owners who did not have the reservoir of experienced local help to draw upon underlined the workers' importance. Edward Larter described weave room help:

> A weaver still has to be a weaver. She's got the responsibility—or he's got the responsibility—of all his looms for the quality, for the production, for the running of the loom. The loom fixer has got to fix them all. . . . Well, the loom fixer takes about a year to be a loom fixer if you're mechanically inclined.
>
> So, all of a sudden instead of having weavers, I'm now having loom starters. There is a lot of difference between a person with two weeks training who knows how to start a stopped loom and a weaver. Big difference. 'Cause I told you it takes me six months to train a weaver.

The fixer was expected to come with abilities to work with machinery, if not looms, thus enabling the mill to take advantage of skills already developed even while it denigrated them as a justification for higher wages. The weaver, too, came with relevant skills and still took months to learn the job, despite claims that the looms had become "automatic."

Furthermore, definable skills represented only part of the mills' need. Edward Stevens described another element:

> You talk about a semi-skilled job, and that's true, but there are a lot of very subtle parts of the skill that make a difference between a successful operator and a non[successful operator]. Any person can learn how to run a knitting machine in a short time, but to be able to spot defects as they're happening, and to be interested enough to do something about it, that's very important too, and to understand the importance of letting someone know that the machine isn't running correctly before it makes a whole lot of bad fabric. All those things.

Lacking a reservoir of skilled labor, these owners attempted to describe the nature of the skills and the importance of the worker.[39] The Lowell approach, to press labor and to devalue skill, had succeeded, but it had not so much eliminated skill as successfully devised a community within which textile skills were inexpensive.

The success of the public relations aspect of Lowell's approach may be seen in the owners' ability to separate the industry's wages from those of the economy as a whole and to avoid a relationship between skill and wage levels. In textiles, even as skills increased, by the managers' testimony, wages remained low and even fell in relation to those in other industries, without regard to the relation of textile skills to those of, say, an auto- or steel-worker. The industry's pattern of operation and its overwhelming dominance of the region permitted it this level of control. Single-industry cities and the divorce of skill from wages were striking characteristics of the industry's techniques in achieving its original goals. Even when it came to recognize the presence of a nonrenewing workforce, the high school graduates of whom the Flathers complained in the forties, the mill encountered few limitations on its ability to enforce this approach and in fact increased the ratio of productivity to labor costs.[40]

Chapter Ten

Operation in the Depression and Wartime

Depression

We have described the elements of cotton mill operation, its products, structures, equipment, and labor, and we can now turn to events of the period. New governmental policy, the unionization movement, in addition to consultant reports and court cases, made relations between management and labor newly visible.

Although 1929 had been a year of great success for the Boott, 1930 and 1931 quickly brought home the result of the Depression, with losses of $450,000 and $273,000. Given the close connection between its products, toweling and clothing material, and the consumer market, nothing else could have been expected. Operations continued at about 75 percent capacity.

By the end of 1930, 120,000 of 280,000 New England textile workers had no jobs, with many more working part-time, making less than $10 per week.

> "There is, perhaps," President McMahon of the Textile Workers declared in 1930, "more destitution and misery and degradation in the mill towns of New England today . . . than anywhere else in the United States." Lowell was economically dead. At this time two-thirds of the labor force was idle; every third store was vacant; doctors could not collect bills; charity was the biggest industry in town.[1]

The problems felt nationwide were intensified in Lowell and other cotton mill towns by severe drops in prices of cotton and clothing, trimming profits (based on percentage) from both ends.

In 1931, the Boott received a $31,600 tax rebate from the city via the Lowell Trust Company just forty-eight hours before the bank closed. On the other hand, the Boott in 1933 remained in a position, along with Locks and Canals, to "relieve" the Union National Bank in a time of need. In fact, it passed dividends only in 1932 (after a 9 percent payout the year before) and 1933. By the fall of 1933, efforts to generate income had led to overproduction nationally, accompanied by further wage-cuts and intensified implementation of the stretch-out. For the rest of the period 1934–52, dividends of 4 percent were standard, with 5 percent in 1936, 1949–50, and 1952.[2]

In the midst of the Depression, government moved to alter some labor-management relations while attempting to calm the troubled economic waters. The National Industrial Recovery Act gave labor the right to organize and to bargain collectively; at the same time it exempted industry from antitrust laws and regulated prices in exchange for a promise of minimum wage and maximum hour agreements and recognition of certain labor rights. George A. Sloan, head of the Cotton-Textile Institute, saw the NIRA "as a means to deal with the overcapacity problem and quickly pushed through the Code of Fair Competition No. 1" in July 1933. That act created a minimum wage of $12 per week in the South, $13 in the North, limited production to two forty-hour shifts, and offered labor an undefined right to fair labor practices. However, the committee supervising implementation in the textile industry interpreted the act to grant the right to bargain only to an organization at a given mill, blocking representation by the United Textile Workers and favoring the formation of company unions.

Industrial Relations Committees were to be set up on the mill, state, and national levels. Since their roles were ill-defined at best, the Cotton Textile National Industrial Relations Board was in position to give shape to the plan. Rather than define standards and mediate disputes evenhandedly, however, its members combined "spinelessness," disinterest, and southern textile industry advocacy. The results spelled disaster for the workers, a "humiliation" for the UTW.[3]

The combination of unemployment, poverty, plus federal legislation which marked the Depression era led to the most active labor protest Lowell had ever seen, and strikes hit the Boott more often after 1930. During the first years of the Depression, wages fell 10 percent on two occasions, in 1932 and 1933. A comparison of Boott rates with the Fall River–New Bedford standard offers a base for comparison throughout the period (see table 9). The increase reflected here led to a flurry of activity on the mill's part to determine Fall River workloads and compare them to the Boott in hopes of counteracting this effect.

Work's highly irregular nature affected worker income. As measured by the percentage of looms in operation during 1932, for example, production fell below 50 percent for four months. During periods of 1933 and 1934, the NIRA's Cotton Textile Code limited production by restricting operation of machinery to 75 percent of available hours or days (not stoppages for full weeks). The Flathers expressed concern

Table 9.
Pay Schedule Adopted by Fall River and New Bedford Textile Councils, July 31, 1933

	Fall River–New Bedford Scale	Boott Mills (40 hours)
Doffers	$13.00	$13.00
Ring Spinners	15.00	14.70–16.44 P[a]
Doffers	16.00	19.04 P
Spoolers (Cotton)	15.00	15.00 P
Winders	16.00 (From Skeins)	15.00 P Foster
Knotter Portable	23.00	24.00 P
Helpers	16.50	20.00 P
Weaving:		
Plain Non-Auto	16.50	—
Auto.	18.00	18.05 P
Fancy Non-Auto	18.50	—
Fancy Box & Auto	19.00 (Dobby)	19.32 P
Jacquard	20.25	18.05 Woven Name
Side Cam	—	20.25 P
Smash Piecer Room-hand	17.00	15.16
Loom Fixers	26.00	23.96
Changers	23.00	20.32

Source: Treasurer's Report, 8/34, box 34, Flather Collection
[a]"P" indicated piecework. These are piecework wages computed as if for a 40-hour week.

that a strike might soon follow the Code's slowdown in 1934, so they made an effort to get orders out.

Labor's new rights and opportunities under the NIRA led to competition among three organizations eager to represent Boott workers. Craft-workers dominated all three: the local Loom-Fixers Club, the Textile Workers Protective Union (TWPU), and the national United Textile Workers, associated with the American Federation of Labor (UTW-AFL). Each represented a different constituency and a conflicting point of view, as the events of 1934 showed. Southern representatives at a national UTW meeting demanded a general strike (employees at the Boott director's eponymous Bemis Bag Mill in Bemis, Tennessee, were among those eager to walk out). McMahon led the effort for industry-wide collective bargaining instead of production quotas, with standardized wages for skilled and unskilled workers, load reductions, and hourly pay increases to offset the effect of dictated reduction in hours of operation.[4] The TWPU opposed the strike, arguing that the goal should be to get a contract, to be "responsible." Flather anticipated that the Loom-Fixers Club would go out "to save face." On the eve of the strike, 2,500 UTW members were expected to stay out and 2,000 members of the Textile Workers Protective Union planned to report. Flather, at this time both Treasurer of the Boott and President of the Merrimack, announced that those mills would open in the morning.

When the general strike began on September 1, 1934, Charette, of the TWPU, sent in 800 workers, but 1,942 unorganized women walked at the Boott. Their absence, along with that of loom-fixers and other key trades of the UTW, prevented operation. Textile workers nationally had struck in what Jim Green calls "one of the most important industrial conflicts in United States history," protesting the National Recovery Administration (NRA) with its "low minimum-wage scales, its acceptance of company unions, its delayed hearings, and its discriminatory rulings."[5]

The Flathers responded by drafting a letter to all employees. They blamed "outside pressure" for the strike, cited the opportunity to "confer freely with management," recalled their support of the hospital, gifts of wood and coal to needy workers, preferential treatment of workers' relatives, and claimed months of work undertaken to tide the hands over when orders were lacking as reasons for a return to work.

Despite its national symbolic significance, the strike had little measurable result in Lowell. The single "largest industrial action in the history of American labor" petered out within a month with no achievement to point to. Management stated it hired back more workers than it felt were needed in order to avoid charges of discrimination and bad feelings. The workers had closed down the industry but lacked the strength amidst economic deprivation even to obtain recognition in Lowell.[6]

Relationships between management and labor were not yet to change. As in the past, dealings generally took place in confrontations between the office and a small, skilled, loosely organized group. Flather described to his board how management had worked to intimidate labor through visits to employees' homes, local contacts, evasion, and a "policy . . . to get the labor leaders out on a limb."[7]

While the economic storms raged across the nation, Flather informed the directors that the Boott was not always part of those trends. When Textile Code Authority (TCA) figures showed a 22-percent decrease in textile production by comparing the first quarters of 1934 and 1935, the Boott's record showed a 66-percent increase, from 3,318,000 yards to 5,531,000 yards. Seventy-one mills closed during this period, but Boott earnings improved. Flather characterized the mill's relationship to the broader situation: "We take the job of getting along with the New Deal seriously, but by 'tending closely to our own knitting' we feel justified in looking forward with confidence."[8]

By the 1930s the nature of the textile industry stood widely recognized. *Textile World* in 1930 lamented the difficulty of restricting production. Harold H. Young found cooperation to reduce production "encouraging," since the "lack of cooperation has been the cause of some of the evils from which the industry is suffering." Similarly, A. W. Benoit believed that in most years a single-shift operation throughout the industry would keep all the mills busy, but he felt no optimism regarding the possibility of achieving such a limitation. A 1934 study found that "the cotton textile industry has been and is desperately sick," unorganized and dispersed, and plagued by "ruinous" overcapacity and a "torturous" distribution system for goods. Benjamin Gosset, president of the southern ACMA, admitted that "it must be evident to all that there is something wrong about our present system, . . . that it is dissipating our energies and bleeding our plants to death." Flather characterized the behavior of two of his larger competitors in 1934, when Nashua Manufacturing was losing money and B. B. and R. Knight liquidated: "[It] shows that people have been making money out of their inventories and they are getting pretty sick of it. Everybody is getting uneasy in cotton manufacturing. You are going to have strikes, the processing tax continued, and the amount of business you can get is being shaken down by one limitation or another." He was observing that companies were taking money out of their businesses in any way available, not just from manufacturing profits, and once that ability ceased, they disappeared.[9] Scranton describes staple textiles as revealing the "near-perfect competition that the entire twentieth century oligopolistic mechanism was designed to evade."[10]

Labor continued to seek to diminish its vulnerability in the midst of these chaotic forces. In 1935, the National Labor Relations Act (the Wagner Act) gave workers the rights to collective bargaining, free speech, and free elections of union representation, as well as the rights to protest unfair labor practices and to seek their redress. Although business tied it up in court for two years, it encouraged the organized labor activity it ultimately legalized. Following the founding of the Committee for Industrial Organization (CIO) in 1935, in part as a response to events such as those of 1934, workers welcomed the formation of the Textile Workers Organizing Committee (TWOC-CIO) in 1937, which became the Textile Workers Union of America (TWUA-CIO) in 1939. Among other results, these organizations had the potential to significantly alter the situation in textiles by giving women workers a voice, an opportunity for power long denied them by the structure of industry and the craft orientation of the labor movement, which effectively barred them from most earlier unions.[11]

In December 1936, the Boott granted a 10-percent increase to all workers, in line with an agreement worked out by New Bedford and other mills. It protested the increase to fellow-managers. At the start of 1937, Flather, having given up his role at and income from the Merrimack, received $20,000 per year. His sons' salaries advanced from $7,200 to $8,700 per year. In January, the State Board of Conciliation and Arbitration (SBCA) awarded the loom-fixers an 8 1/2-percent raise, which the mill appealed. In March, the Boott agreed to another 10-percent increase, again in line with other New England mills, although "deploring their action."[12]

In early May 1937, the loom-fixers walked out. The next day a program of picketing and CIO sound trucks combined to draw out 1,600 employees. Eight days later, the CIO contacted the office for the first time and asked for a 15-percent wage increase and union recognition. Management resisted; it cited the two 10-percent raises given since December, expressed willingness to study equalizing wage differences between jobs, claimed that the negotiations represented de facto recognition, and demanded a return to work. It also refused to sign the former Cotton Textile Code, in effect a Code of Ethics, as "too dangerous." Management then claimed that the Greeks employed violence to intimidate the French (generally reluctant strikers) and sought an injunction. Board of Arbitration intervention led to an end of the strike after forty days, with no raise, no discrimination in rehiring, arbitration in case of discrimination or in case of a wage dispute, a Grievance Committee to be established, the TWOC to be recognized as bargaining agent, the forty-hour week continued, and no strike or lock-out. The six-week strike achieved the significant goal of union recognition, a fact Flather omitted in his report to his board.[13]

This settlement held until October, when the doffers went out for two days. The State Board got them to return on the basis that they would receive the workload and wages to be agreed upon in the concurrent Fall River–New Bedford dispute. A year which had seen conflicts with loom-fixers, the entire workforce, and finally the spinners and doffers had finally ended with few clear accomplishments for any party but with bad feeling widely shared. Regarding tactics during the strike: "We didn't carry out one of Mr. Roberts' [a Director] favorite plans, which was ostensibly to go South. We may go yet, and not ostensibly. Whenever you are ready to give us $500,000, Frederick is ready to show you where we can make more money than we can by remaining here."[14] By this time, the board had decided to take money from the Boott Mills as long as it produced but not to attempt to continue the mill through further investment in plant or equipment. As Paul F. McGouldrick pointed out, "The durability of textile plant characteristic of industry technology meant that its owners could extract quasi-rents for decades rather than years." While directed at nineteenth-century operations, his comment described the Boott in both centuries.[15]

Immediately after the 1937 strike, Flather noted: "We naturally expected that the Strike would set us back financially in both May and June, and of course are gratified that it did not." Flather formed his views of organized labor during his years at the machine shops and International Harvester, before he encountered a union in Lowell. His assessment of the CIO in 1937 revealed his feelings: "C.I.O. Lewis has gained for them [steel workers] eight hours of *idle time* per week. . . . 20 cents less, eight hours+ to spend. . . . They probably had enough *idle time* in which to spend more money than they had as it was and did not need the eight hours more of *idle time* with less to spend" [emphasis added]. For him, time for workers represented idleness, not freedom, discretionary time, or anything else positive. He later described Lewis's plan as a "Coming Labor Upheaval" which would make him "Big Chief of Everything."[16]

The year 1937 saw another economic low-point: "Flather: 'I don't believe we have ever had so rotten a depression as this one is. People are taking the other people's business.' Roberts: 'You have no such depression as in 1920.'" Mills behaved as competitors rather than colleagues, a recurrent, but unwelcome, situation.[17]

Wages fell 12 1/2 percent, again in pace with Fall River–New Bedford, or $2 to $3 a week, in 1938. When most New England wages rose 7 percent in November 1939, the Boott refused to follow. The Boott rode out with better than typical success the downturn of 1938–39, in part by operating with just 593 employees in May 1938. Only in 1939 did employment exceed 1931 levels, at 872, as they began to move into war production. In 1940 the Boott employed 1,132.

Ownership recognized the effects of low wages: "The people have been living from hand to mouth for eighteen months, and now that war has broken out, prices have shot up in almost every industry," observed Director Roberts in 1939. Flather could even agree with the desirability of a forty-eight-hour week and similar laws, but only when all New England and, ultimately, all the United States adopted them.[18]

Wartime

As time passed, labor's opportunity for organization decreased. In 1939 the Supreme Court diminished the Wagner Act, ruling that the Act did not interfere with the employer's right to hire and fire; the Court permitted the employer to impose terms in case of an impasse and blocked strikes during the term of a contract; it prevented workers from seizing a plant; and it imposed on the union the responsibility to police its members and insure responsible behavior.

In November 1940, the loom-fixers struck in what would prove to be the last craft organization effort at the mill. This group, French-Canadian with just four exceptions, went out despite lacking, in management's opinion, the support of the Franco community. The mill did not close and kept the looms running with the help of supervisors and weavers.[19] This dispute stemmed from the demand in 1939 for a 7-percent pay increase, spent many months in arbitration, and was only settled when in April 1942, the state courts ruled that the arbitrators had erred in failing to consider both workload and pay in their decision for the fixers. Management had spent a large amount of money in arbitration in an effort to block organization. Already involved in war production, the Boott complained to the Navy that it had lost many skilled loom-fixers and skilled weavers, some of whom didn't return, and that others left after the strike. National unemployment rates of 14.5 percent of the workforce in 1940 and 9.9 percent in 1941 helped explain the corporation's ability to find any replacement workers.[20] Still, only by paying premiums and bonuses could they keep weavers to maintain production. All this led them to ask the Navy for $109,122 in compensation for unexpected costs.[21]

By 1941 the War Labor Board added to the Supreme Court limitations a ban on strikes and walk-outs, with compulsory arbitration. Wages rose 10 percent at the Boott in 1941, but at the same time prices and profits escalated rapidly. Executive salaries also rose, and the two "boys" began to receive $15,000 per year as of January 1942, double their pay six years earlier.[22] The pay of wage-earners had risen about 20 percent during the same period. In 1942, the WLB ruled that the only allowable pay increases would be for cost of living above the level of January 1941. Giving a modified closed-shop arrangement to labor in return scarcely shifted the balance of power, which remained with the owners through the war.[23]

The Flathers were in turn manipulative, concerned, condescending, perplexed, and mistrustful. When wages were to be raised, they consulted with their "Senior French Canadian Supervisor" in order to make the timing most effective. They employed "women in as many operations as possible, as being easier to train and more stable."[24] Management recognized the connection between the wages they offered and their ability to attract, motivate, and keep workers, particularly the ones they wanted. John Rogers argued that they would "have to pay more" to achieve their ends, a voice against savings at any cost.

A. G. Cumnock. Courtesy the University of Massachusetts, Lowell.

Frederick A. Flather. Bachrach Photograph, courtesy the Flather Collection, Lowell Museum.

John Rogers Flather. Bachrach Photograph, courtesy the Flather Collection, Lowell Museum.

Frederick Flather. Bachrach Photograph, courtesy the Estate of Frederick Flather.

When a new figure entered the picture in the early forties, the Flathers saw him in old terms:

> A neighboring cotton mill fell victim to an ambitious independent local Greek organizer, Louis Vergados. He first usurped the throne of the old Loomfixers' Club, fired its old-time officials, flattered the French making [Herbert] Valencourt [sic] President of the Loomfixers' Club, but eclipsed that Club by a greater new organization, patterned after the tricky shoe shop organizations (which found fertile ground in the fly-by-night shoe shops), and is called the "Lowell Textile Independent Union," Louis Vergados, Organizer.

They saw flattery and trickery as the keys to labor's moves, and they saw workers as subject to manipulation by the likes of Vergados.[25]

Despite efforts on the part of management and the *Courier-Citizen* to minimize Vergados's impact, his role grew. The union distributed this appeal in 1941:

> Attention! Workers of the Boott Mills; Over 65 percent of the workers of the Boott Mills have already joined The Lowell Textile Independent Union:
>
> 1. The United States Government gives you the right to organize and become members of a UNION.
> 2. The United States Government gives you the right to choose who you want to represent you.
> 3. If you want higher rates; better conditions; one week vacation with pay; elimination of the stretch out or overload; and to stop discrimination by the supervisors, then join now.
> 4. So why not join now with the LOWELL TEXTILE INDEPENDENT UNION and derive the benefits of a UNION contract that are being received by the workers on the Merrimack Mills; where the LOWELL TEXTILE INDEPENDENT UNION has a Union contract, representing all the workers of that mill.
> 5. Our offices are located at 373 Moody St. and are open from 9:00 A.M. to 5:00 P.M.
>
> JOIN NOW AND FULFILL YOUR AMERICAN RIGHT; Fraternally yours, Louis Vergados, Organizer; LOWELL TEXTILE INDEPENDENT UNION.[26]

The CIO also kept up its organizational pressure and distributed leaflets, newspapers, and flyers throughout the area. James W. Bamford, from the Lawrence CIO office, claimed a majority of Boott workers in August, while Boutselis, of the Independent, countered with assurances of Greek and French support for that organization. In an apparent deal the next month, the two unions stopped competing and divided the prize: the CIO dropped its drive at the Merrimack, as did the Independent at the Boott.

A cross-check of signed union cards with the plant's payroll led to the recognition of the TWUA-CIO as labor's representative in October 1942. Ensuing meetings brought together management and the union representatives, led by President Vaillancourt, to discuss contract proposals. As negotiations progressed, George Baldanzi, Senior Vice President of the TWUA, came to play a role.[27]

Boott management repeatedly turned discussions with Bamford to the advantages of southern manufacturers, and he made a good impression on them with his understanding responses. He offered the services of the CIO research department head, Solomon Barkin, in case of workload disputes, to enhance Boott competitiveness. His assurances of assistance with such problems led the treasurer to consider him "a square shooter," and John Rogers to label him "an enlightened labor leader."[28]

By the time of the War Production Board's (WPB) 1942 order restricting the purposes of textile manufacture, the Boott had already shifted to nearly complete concentration on military contracts and ran three full shifts starting in 1942. In that year the Army and Navy awarded the mill their "E" production award to recognize the achievement of the men and women working there.[29]

Negotiations with the union within the restrictions of the War Labor Board regulations led to a contract including a Management-Labor Committee (MLC) drawn from each department (Carding, Spinning, Weaving, and Finishing) and each shift, a grievance procedure available to each side, either a week's vacation or a week's pay (the employer's choice) after a year's employment, a forty-hour week with time-and-a-half for overtime, no strike or lockout, lay-off by seniority and ability, dues deduction on individual authority, and allowance for veterans' return after the war. Wartime restrictions and profits made it easy for management to accept this undemanding contract. They warned supervisors to remain neutral on all issues regarding union membership and informed them they would be discharged for discrimination. The Boott attempted to turn the situation to its own advantage when it got permission from the union to print the contract booklet for the employees in larger and more readable type than usual, with a picture of the mill's "E" pennant of the front and a picture of the award ceremony included, in order to "make it quite an outstanding booklet to the recipients."[30]

The first year of union-management relations, 1943, passed calmly. Pirating of workers by local defense plants continued to exercise management more than most union issues. The MLC, or shop committee, worked to improve production, increase worker earnings, and enhance the prospect of the mill's survival: something for everyone. After government's complaints about imperfections, particularly slubs, in the uniform cloth, the committee visited the Naval Clothing Department (NCD) in Brooklyn. Correcting the problem meant slowing production in the carding department. The MLC proposed a reclassification of carding jobs with increased pay for the newly defined tasks. The reduced speed of card-room operations led to decreased production on down the line, and rather than lay off the loom-fixers who were affected, it was decided to reassign them temporarily. Management glossed over cases of minor disagreements in order to avoid trouble during these profitable times. Initial attempts at cooperation demonstrated a potential for future development and the capacity for the two sides to work together. Remaining tensions led Fanning, the Boott's efficiency engineer, to tell the committee's union members to ignore any "remarks, jibes, or slurs" directed at them by other workers for their participation.[31]

Issues of workloads remained a central point of discussion. Bamford indicated the union perspective in 1943 when he cited problems of layout and flow which would have to be treated before any assignment increases in order to avoid making employees think they were expected to work harder. He saw greater loads flowing from improved conditions, not greater exertion. The MLC agreed to treat, in order, job descriptions, job assignments, job efficiency, and, having settled those issues, earnings for them. Both sides conducted time studies in an effort to determine acceptable workloads. They accepted the approach to job assessment, but they differed over its use. Since the 1820s textile mills had been utilizing elements of what came to be known as "scientific management" when Frederick W. Taylor attempted to apply it to machine shop practice around the turn of the century. Dividing jobs into their constituent parts, assigning minimal portions to each employee, and paying piecework wages for workers and overseers, for example, had long histories in textiles. Time studies attempted to enhance these techniques and to justify further subdivision and intensification of the labor process. The management time-study engineer, W. A. Hawkes, made weave-room studies "based on conditions that he expects the Union to help achieve here by better loom fixing and better methods on the part of workers, etc." Toby E. Mendes, the CIO time-study man, dealt with actual conditions there and recommended new bobbins, sharper feelers, new uniform pick clocks, cleaned warp stop motions, and even "increased air hose pressure to remove grease and dirt from the looms." In other words, he expected the mill to put itself in order before looking to the workers as the source of improved operations. Continuing the same dispute, W. Malcolm, union vice-president, complained of the poor quality cotton being processed and "wants pick sheets [identifying workers' production each week] down," while Fanning said "competition among room and weavers was a good thing." Since this confrontation took place during wartime, it ended up before the U.S. Conciliation Service, where the company accepted "the bulk of the Unions' work assignments clause," with Hawkes's assurance he could fulfill his purpose within that context.[32]

Although the Flathers recognized the limitations placed on operatives by the machinery, buildings, and conditions, they added the union as a factor "that enters into the restrictions placed upon us." Claiming that tradition blocked increased loads, they even cited Loper's study, which had actually blamed not tradition, but the mill. Individual workers, however, those working in the proverbial trenches, rejected even increases acceptable to the union. "No more sides at any price," was the spinners' reaction to an offer of increased pay for a temporarily increased load to compensate for the mill's inability to hire enough spinners. Only when management pleaded with them individually, a unilateral act its own lawyers warned was illegal, could it cajole acceptance of its request.[33]

In 1943 a description suggested something of the nature of the work, its difficulties, and the difference in point of view between management and an employee on piecework:

> Results of spending 45 minutes with E. Ntapalis, a spinner, who works in 6–5 running eight sides of No. 15 combed warp. . . . During this period she had to piece up 64 ends and put in 26 new rovings. This is equivalent to 120 operations per hour. It was noticed that whenever she put in a new roving she would attach the end of the roving to the roving of the second bobbin and break off the roving that was running out. Each time she did this there would be an average of three or four inches where 3 strands of roving were being drafted instead of 2. It was also noticed that at times apparently the traveler would be sticking and the spinner would touch her finger to one of the bearings getting a small quantity of oil on the tip of her finger, transferring it to the traveler and then piecing up. This would give you an oily place in the yarn. . . . Mr. Lemire, supervisor of this room, said this was a very good spinner. In talking with Mr. Bourgeault about it, he said that every time a piecing was made we would get a slub.[34]

Management's description ignored several factors. Ntapalis repaired 175 percent of the breakage calculated for her job, worked with newcomers paid equally for less production, suffered from poor work preparatory to spinning, from travelers which stuck, and had neither opportunity nor incentive to overcome problems ranging from dirty cotton to weak roving. The need to learn from Bourgeault that Ntapalis's procedure produced a slub does not indicate processing expertise in management.

The argument was basically circular. Spinners were assured that the quality of roving coming to them would soon improve, permitting better earnings, while the union's Malcolm asserted that "until the yarn was better there was no need checking" individual efficiency. The next year, 1944, Flather informed that "he [Frederick] is getting along with the introduction of all waste for toweling. The only trouble is, spinners who have been working on easier-going long staple do not like it. . . . In time they will naturally be re-adjusted." He attempted to deny the connection between quality of fiber and size of load. Barkin referred to Mendes's report regarding poor conditions, but management cited their consultants', ignoring the fact that those also complained of poor operating conditions. Workers objected that they lacked combed cotton, lacked bobbins in the weave room, and lost money because of elements outside their control. The Boott's own efficiency experts described the effect of conditions on creeler and warper tenders: "The amount of walking to find the ends at the creel is a great deal more than credited in the time standard. To consider this a piece rate job seems more or less foolish in light of the operating conditions."[35] Over time and throughout the mill, the refrain continued.

Relations soon became more tense when both sides began jousting for position in contract negotiations. The union wanted the Boott to accept the Fall River–New Bedford agreement as its contract, a move the mill claimed would cost it $480,000. According to management's account, the union offered a strong opinion on the basic nature of its relationship with the Boott:

> Mr. Barkin said our economic claims of losses was not the Union's fault, that it was Management's problem, that the Union was not involved, and that they could not ask the employees to give up some rights because of Management's problems. . . . [He] said that our average [wage] was *lower than the average in the South,* and we said we hoped the South would be higher, [that we] would be able to survive. He said he thought our average was about 57 cents an hour judging from some wage figures he had gotten from us during negotiations on December 15, 1942. We explained that southern mills have a double advantage, steady runs of one cloth per mill. [emphasis added]

Management believed southern wages equalled 91 percent of the Boott's. Barkin sought fair pay for honest effort on labor's part. He accepted no responsibility for difficulties in the operation since the workers had no power to change conditions or circumstances, only to provide effort, which they did. Management saw itself as the perpetual victim of the unfair advantages of others. It welcomed his suggestion that average pay lay below that of southern competitors.[36]

Workers reacted to actions they saw as infringing on their rights and their dignity. In one instance, winders and spoolers walked out because

Morin, a nonunion employee, was fixing and scouring their frames. Ralph Motard, TWUA Chief Executive for the Lowell District, supported management, recognizing the limitations imposed by the lack of a closed shop. In another instance, a returning worker objected that she was not given the identical frames she ran before a layoff, reminiscent of Mary Metcalf's complaint in 1845 and the notion of "factory culture" Scranton notes in the period before 1880. While management contended the move was made to reduce her "talking with a chum," the union insisted on her position. Workers felt a new status and opportunity to express themselves, and their organization played a new role in arbitrating their position on innumerable issues. In many, even most, cases, what it required of management was an appearance of fairness, an observation of minor worker interests.[37]

In 1945 the Boott presented its case against continued operation in Lowell in a court appeal for a tax reduction.[38] It revealed both conditions and management attitudes. They discussed future production only in terms of "present machinery," displaying their determination not to invest in equipment. They complained about producing in a "marginal area" and admitted that producing a variety of styles taxed conditions at the Boott. Old, weak, and deteriorating buildings limited labor efficiency more than any legislated "labor restrictions" cited and contributed to a less than enthusiastic attitude by the workers. The fact that they did not intend to modernize equipment or plant, had not followed such a policy in the past, and did not move to eliminate "layout restrictions" defined management's position. That workers avoided the Boott represented labor's response.

During the preliminary stages of the negotiations which would ultimately lead to a new contract, the union pointed out its potential to help the mill with its problems. Motard cited his work at another Lowell company, Hub Hosiery, in instituting necessary wage reductions to keep the operation going. He implied he could do the same at the Boott in exchange for a closed shop. However, Barkin was finding the situation difficult: "He said it was important also that we overcome the hurdle of contract negotiations, and he is sorry that we were not prepared to [begin] serious practical negotiations. 'It is imperative that your attitudes really change to permit a cooperative basis for solving your many problems.'"[39] Meetings took place most often at Lowell, although occasionally they moved to the Boston office of either side. The Boott management suggested only men attend when meeting at their Boston office, for some reason, and implied surprise when it visited the union's: "Offices were dignified and businesslike, neat, not gaudy."

The TWUA worked from set policies and procedures. President Emil

Rieve had established the Fall River rate as the basis for all negotiations in New England, a position endorsed by the War Labor Board. It had notified twenty-five mills in New England, twenty-three in the South, and five in New York and Pennsylvania that they would pay a 55-cent-per-hour minimum, pay occupational minimums known as "peg points" to provide the usual spread between various jobs (keeping the various wages at the proper level above the minimum), and increase wages 5 cents per hour for all earning over 50 cents per hour. The Boott had to comply to avoid a strike and planned to seek a price increase of 1/2 to 1 cent to cover the expense.

The local committee for the Boott could discuss job descriptions and loads related to the mill but not issues of wages. It saw a closed shop as essential to any talks, and the Boott reluctantly agreed to discuss the concept. The MLC gradually agreed on job descriptions through the mill, at times noting process failures on the way (bad bunches being built on the spinning frames, for example).

In May 1945, the TWUA presented its overall demands for a 65-cent-per-hour minimum wage, a 10-cent-per-hour increase, shift differentials, one week paid vacation after one year, two weeks after five, paid health insurance, six paid holidays, a half-hour lunch break with pay, severance pay for jobs lost to technological improvements, a union shop, and preferential hiring. The two-year contract proposal also included daily minimums for piecework with the average employee making above the guarantee.

Negotiations over these items coincided with WLB hearings on workloads and rates for New England mills. At the WLB, the union made it clear it anticipated only minor variations from the Fall River standard previously endorsed by the WLB directive. When Barkin noted that after three letters he couldn't understand the Boott's meaning in saying they paid piecework and hourly workers "as if" they were doing full assignments, John Rogers interpreted the statement as an endorsement of the mill, not a query about its impenetrable logic. The union (Barkin) expressed its basic position: the sixth worker (of ten) had to be able to earn above the average or guarantee; in other words, 60 percent of the employees would be at least at that level. Given that, it could be flexible about assignments, as it had been in Paterson, New Jersey. However, it could not negotiate with a mill which could not compete while paying the Fall River scale. The TWUA vice president, Baldanzi, expressed the underlying rationale for the union's position: "The mills had made so much money during the war that they could afford to operate at a loss for some time to come." The time had come for the workers to share in the wealth they had helped the companies earn while their wages had remained controlled. The union believed that available funds could produce these raises without a price increase for the product.

Boott management, however, quoted the president of General Motors: "General Motors cannot and will not use money saved up for many years to modernize and expand its plants and for providing more good jobs, to pay excess wages for work not performed." The quote was irrelevant, for the Boott could not legitimately pretend to plan "to modernize and expand." Only increases in production could lead to increased wages in such a view; any other raises appeared to be "excess." John Rogers also cited literature advising that it was "unwise to distribute all the savings" of new technology, "as that would equal a tax." New machinery, he felt, handicapped the unionized section of industry, presumably because the owners might be less able to reserve its savings for themselves. The Boott's equipment made the issue academic. In any case, the worker was placed in a double bind if he or she could only expect raises through increased production, but owners reserved technology-based production increases and their savings for themselves. The implication was that the workers could only increase wages by working endlessly harder.[40]

The negotiations ultimately failed and led to a strike. Discussions turned on a trade-off between workloads and a closed shop. According to management, the union, through Bishop, indicated that the "Boott ought to be entitled to the same amount of production that other manufacturers are getting on each job, qualified only by conditions peculiar to the individual mill, as there is no doubt that the Boott has a unique special problem." On the other hand, these "conditions" referred to may have been the ones limiting production which consultants had so often noted. He added the caveat that only with a 100-percent guarantee against layoffs stemming from increased loads would the union help institute the revisions under discussion. Given the satisfaction of all his conditions, he agreed to help get the workers to accept the loads management bargained for. The Boott joined Whittenton, Naumkeag, and Nashua Manufacturing to form a coarse group within but separated from the Fall River–New Bedford orbit, but the effort soon fell apart. The strike ended October 29, with the announcement of an agreement coming on November 19, 1945. The union had gotten its minimum wage, an 8-cent-per-hour raise, vacations, third shift premium, reporting pay on each shift, and a closed shop. Management expected union help instituting new workloads.[41] The contract offered mixed results, creating as many issues as it resolved. In fact, the Boott soon decided it had given more than some other mills and worried that it couldn't live with the contract.

The contract did not ease relations at the Boott. The union reiterated the position that there would be no help on workloads without assurances against layoffs and cited loom-fixers put out of work as an example of the problem. The mill would only promise to work at avoiding unemployment based on increased loads, leading Brown to observe "that it appeared the management was incompetent, and that the mill should be turned over to the employees, and then there would be no capitalistic profits to distribute," a point of view not likely to win favor with the Flathers or their directors. This impasse continued for months as the union maintained its position and the management felt as though it was being victimized by this adherence to a policy of ensuring jobs.[42]

Consultants had continually indicated that loads could not be increased without improvement in conditions. Compounding the problem, the loss of skilled employees to better jobs introduced people who could not always work up to performance standards; in order to keep any help, the mill had to pay them a minimum wage. Experienced workers then had little incentive to work industriously or efficiently, since to do so resulted in minimal income difference over those whose work did not meet standards. These problems further complicated the historically troublesome disagreement over workloads.

According to Hawks, rates for jobs were calculated based on an assumption that the machine and operator could expect to attain 85 percent or more efficiency of operation. Therefore, "any improvement in efficiency up to 85 percent is management's saving, not to be shared with the employee." As in the case of technological improvements, only the owners were to gain from increases short of his arbitrary definition. During the forties, management made constant efforts to raise the workloads to those of the Fall River–New Bedford standard. In part because it wanted to make these increases without raises in pay, workers consistently resisted. Management claimed that its demands did not represent harder work: "The newer equipment has attachments which make it easier for the machine tender to tend more looms without putting in any extra effort . . . , but we have difficulty with them in that respect." Such a view left no room for consideration of issues related to conditions in the mill where antiquated equipment and oppressive conditions made 85 percent efficiency a distant ideal, and where workers did not operate "newer equipment." New work techniques were seen by management as sources of labor-cost savings, and time-study as a defense against claims of arbitrary or unfair demands. Though wages would not be increased until the factory's production rose beyond a predetermined standard, management claimed that income would rise from steadier operation of the mill as its competitive position improved and it secured more orders. Later in the same document management acknowledged that the record did not bear out this assertion.[43]

Disputes continued over efforts to increase workloads in the spinner-doffer job. Angelina (a.k.a. Emma) Belisle was first-shift union steward and a spinner on 6-4. She participated in committee meetings with various supervisors both on the floor and in the counting house and performed, and argued about, the work under discussion.[44]

When management wanted to raise workloads in spinning in 1945, it encountered difficulties from a joint advisory committee on which Belisle served:

> It was brought out that Mrs. Belisle's present job, 16 frames, kept her working approximately 21 minutes out of each hour, and that all we were asking her and the other doffers to do was to work approximately 35 minutes out of each hour, or 1,400 minutes per week, out of 2,400, leaving the doffers with 1,000 minutes, or 16 hours and 40 minutes per week idle time. . . . The trend in Mrs. Belisle's mind seemed to be that she wanted the doffers to work about one hour and then be permitted to leave the mill for an hour or more, and still be paid full time.

While one would like another version of the meeting, the conflict was clear. Doffers were removing and replacing about 500 bobbins per hour. When the committee moved through the spinning rooms to view the work, some workers were waiting for frames to fill up and were seen eating lunches or reading newspapers. Performance was less than was specified by a 1937 Board of Conciliation and Arbitration agreement. However, one finds here no mention of the problems cited by consultants and workers which made workloads and earnings hard to predict: changing yarn counts, transportation delays, poor or poorly prepared stock, varying machine speeds, etc. The doffers resisted being pushed toward doubling their performance without an improvement in conditions or an increase in pay. Union officials complained of mismanagement, shortages of roving and bobbins, and a lack of expertise at the top. The Flathers routinely dismissed or ignored criticism of management and material.

The dispute continued into 1946. The engineering department described problems observed as the fault of the fixer, roll picker, or spinner. Some workers managed to keep their frames running while others allowed several frames to stop with full bobbins and did not keep up with the work. Problems, from management's viewpoint, stemmed from labor's unwillingness to do sufficient work and the union's refusal to enforce the mill's desires. The conflict in point of view was as old as the mill itself. The advent of the union had given workers both strength and a voice with which to protest expectations they found unreasonable. On the other hand, both Agent Thomas's circa-1910 statement and the Flathers' descriptions suggest that the workers cooperated with changes they found reasonable. By January 1946, the union agreed to an assignment of 450 bobbins per hour at current pay, with one cent for every 41 bobbins beyond that. Doffers demanded that trucks of bobbins be delivered closer and that empty trucks be available closer. Significantly, Charles Metcalf had complained in 1844 about the exhaustion resulting from delivering and stacking 5,600 bobbins of roving in a nearly 70-hour week while trucking away another 5,600. Belisle and her coworkers agreed to doff, piece-up, and replace 18,000 bobbins per 40-hour week, an enormous increase in the intensity of labor unaffected by any technological, labor-saving, development.

Loom-fixing presented parallel disputes. Union representatives and men in the mill agreed that the prewar set for a loom-fixer was 62 looms, and that many in wartime tended only 48–50. When some fixers agree to move to 72-loom sets, the union blocked the change. Later in 1945, however, they accepted it on toweling, "on the basis of conditions as existed today." This standard, carefully couched, still gave the mill a performance higher than the prewar standard despite the declining condition of the looms, which the fixers noted.

In 1945 Tully, in the tax abatement case, testified that the looms were not only divided between ten rooms on three floors, but "I believe that the looms are in need of reconditioning. They appear to have been neglected and not properly maintained." Southern mills ran efficient looms, he stated, and loom-fixers could more easily care for 100 good looms than 60–72 not in efficient condition. John Rogers had also noted the layout's interference with this work. The loom-fixers' response to company pressure was particularly interesting. They emphatically rejected the idea that their wages should be set according to those available in the South, on the basis that they "can't be expected to live like" southerners, "can't live on the money we earn now," which they characterized as "starvation wages," and "are not going to try to compete with the South." Rather, they expected to be paid "according to the wage scale in this part of the country." An interesting comparison of the standards of living and costs, North and South, emerges from the account in *Like a Family* of southern workers responding to wage increases by taking more time off. Boott workers had no such luxury.[45]

Still, management felt mistreated, believing that by paying the mandated pay-scale, "we were buying cooperative efforts to improve work performance." Actually, cooperative they got, but not acquiescent, as they seem to have hoped. Loads increased consistently and the feared union intransigence, as opposed to careful scrutiny, did not appear. Despite the impression they gave, the managers had been getting what they wanted all along. The goal, from the start, had been to reduce the relative number and cost of employees needed to produce a given amount of cloth and profit. As early as the spring of 1946, reductions in working hours and increases in income canceled out one another. Moreover, Boott management noted that whereas textile wages had equalled those in steel a generation before, because of southern pressure they were no longer close, "to everyone's detriment." Low wages did not result from a lack of awareness of the workers' plight, but neither did management find a need to share the blame or improve the situation.[46]

When a veteran returned home, management complained that "his family (our employees) takes time off to help him spend his money."[47] Such distrust justified management behavior, whether opposition to the closed shop or fiddling with the books to hide income from union negotiators. It led to individual checks on all unemployment claims (over 90 percent of which were upheld) and established the point of view from which union representatives and their actions would be seen: "The New York high-priced Union officer, Solomon Barkin, who dominates any hearing from the U.S. Senate down, his companion Emil Rieve, National War Labor Board member, and Mr. Wyle, their shrewd counsel." These were people to be duped, avoided, or defeated, some of whom were "even more of a radical" than others, but who as a group demanded "socialistic fringe payments" to workers and "interfered" with the relationship between them and management. These people appeared determined to destroy the very things that management stood for and that had made it rich.[48]

Management was duplicitous in a manner at once self-serving and self-condemning, upon reflection, when it claimed in the 1945 tax case that its difficulties in attracting labor to the mills dated only from the immediate period, that previously French-Canadians, Greeks, and Poles "not accustomed to this American way of life" had not complained or hesitated to come to work there, that only the new American-born high school graduates had raised the issue, made it a problem. Claiming that the lack of elevators, poor washrooms, and the contrast to single-floor well-lit plants had not been an issue previously denied the existence of contrary findings by every consultant the mill had hired for over forty years. It also revealed a willingness to take advantage, to aim for no standard of decency not absolutely required, to treat the workers as poorly as their level of deprivation might force them to accept.[49]

The issue of bonuses and discretionary wages further exemplified the conflict in attitude between labor and management. While management saw union demands as "socialistic," the union might describe bonuses as undistributed wages. Thus management had the wisdom to assign to the widow of Stanley S. Kent, late chief engineer of the Proprietors of Locks and Canals (not the Boott), $7,500 as their share of his death benefits, and to the widow of Nyberg, superintendent of Slashing, $250, and to a fifty-year office employee, a gift. Discretionary kindness could also be seen in efforts to extend the practice of favoring supervisors' sons in hiring to pursuing a spot in the machine shop for the son of local union head, Motard, and finding an office job for the daughter of a worker who translated for John Rogers during a spontaneous walkout and encouraged acceptance of his settlement offer. All the benefits of the operation belonged to management to distribute as it saw fit, or saw helpful.[50]

During the war, the mill lost many workers to better paying jobs. After the war, new problems emerged, according to the treasurer:

> Those employees who left for "cost-plus" munition works formed bad habits regarding easy money and soldiering on the job, and are now preferring "The Nine O'clock $20 Club" tax free per week, to minimum pay of $26 per week, which less about 10 percent withholding tax, is $23.40. They reason that to get $3.40 per week, they would have to work 40 hours.[51]

The slight difference between unemployment compensation and Boott Mill earnings underlined the low level of the wages.

In 1946 labor received an 8-cent advance awarded by the arbitrators in an earlier dispute. Later in the year the CIO sought a 15-cent increase, but despite a drop in the price of cotton, the mill resisted and the union received only a 10-cent raise, which increased the minimum wage to 83 cents per hour or $1,726.40 for a year of full employment. Management admitted that the raise in wages and the drop in cotton prices "offset each other in effect on selling prices for yardage, but a lessening demand for goods would affect prices." However, President Truman vetoed the renewal of the Office of Price Administration, allowing prices to soar while wages only crept ahead. The Boott raised its prices 10 percent.[52]

The end of the war in 1945 led to diminished work in 1946. Disputes over spinning assignments continued, and management persisted in its belief that the stock was running well enough (breaking and requiring repair seldom enough) to permit an assignment of an increased number of sides to each operative. John Rogers visited Spinning with a union representative and found few ends down, even while Belisle left her frames to talk with the union man. He also noted that the fixer on the floor, Joe Comtois, was found in an "empty storage room in the adjoining building, . . . sitting on the window sill." Not enough work was going on to please the office, and an effort at arbitration was attempted, in keeping with the contract. However, the arbitrators sent the question back to the mill for a further attempt at changing work loads, with management looking toward 900 bobbins per hour; study resumed. While management felt twelve sides per spinner was the right load, it decided to attempt an increase from eight to ten sides. The union did not object to extending the assignments of fixers, doffers, and sweepers. After carefully improving conditions in Spinning, management attempted to make the change to ten sides by approaching each spinner in one room. Three agreed, but the fourth not only declined but refused to leave. The next said it would "make her nervous," and so it went. Finally,

> Mrs. Belisle, the head steward . . . said the union had instructed the spinners that they were not to take more sides, and that if anyone refused and was suspended or sent out, the other spinners would all sit down and not work at all. She ended by saying, "Let the head of the union come down and straighten it out."

John Rogers's account reflected the mill's new situation in many ways. He acknowledged the cooperation of the union in the elimination of certain positions, and tacitly admitted the problems which had at times plagued the operation in noting the efforts of the supervisors to make sure the work was "running well" when attempting the stretch-out. That the assistant treasurer would participate in this effort indicated the new relationship, the new degree of power on the workers' side. The degree to which the workers were personalized and dealt with separately indicated that a new awareness of their individuality grew out of their collective strength.

Belisle threatened to adopt the sit-down. Her idea that the union representatives should "come down and straighten it out" reflected her use of the new organization. Despite the effort to make the work run well on this day, Belisle feared that the conditions might well change, revert to what she had experienced as normal. The union believed that the change required fuller explanation in order to make the employees agree to try it, and after two explanations by two union representatives, they did accept it. Collective bargaining had fulfilled the long-time fears of Lowell's investors: labor had a voice, an opportunity to interfere formally with production, in contrast to its informal techniques in the past. On the other hand, the union *helped* impose the new assignments, fulfilling Agent Thomas's belief that unions assisted labor-management relations.

The Flathers were aware of the Boott Mill's deficiencies. They noted the lack of repairs and maintenance. They knew the consequences, and they admitted in 1945 that "no one in this situation can build hopes of a longer economic life for the Boott Mills than five years." They were as aware of solutions to the problems faced by the mill as they were of those difficulties. While John Rogers claimed in 1945 that a reasonable disposition for the buildings would be to level them and use the bricks for fill for a "scenic highway," a "marine park," or a "restaurant," at the same time Frederick prepared a modernization plan for the complex. According to his draft, the short narrow rooms, alleys, ramps, stairways, and elevators were responsible for 15–30 percent waste. He proposed roofing over the yard and removing the top three or four stories of the buildings, creating low buildings and big rooms. Substantial savings would then be realized in supervision, transportation, and heating and cooling, along with improved supervision, quality, efficiency, personnel, and taxes. Each month's savings would equal two months' construction costs, prices could be lowered, and the mill would be placed in a position to compete with anyone at a total cost of just $200,000 to $250,000. Not only could such a plan easily have been followed, simply paid for from income or the Net Quick, it required no further demands on labor. In fact, labor savings were not even listed in his rationale. Despite that, the plan was not pursued. It remains interesting for its specificity and its assumption that the labor situation in no way precluded it.

Conclusion

The Boott continued to pursue new products it could make and market profitably. During this period curtains, corduroy, toweling, and uniform cloth represented the staples around which production revolved. Finishing added an income source and improved flexibility.

Yet labor was slow to benefit, and trends continued to heighten its risk. Only in 1944 did employment become sufficiently regular to produce a lowered rate of taxation by the Massachusetts Employment Security Tax.[53] When the City of Lowell proposed to raise the taxes on this profitable plant in 1944, numerous studies poured forth showing the disadvantages the mill faced on all sides. Between 1929 and 1945 the ratio of active cotton spindles, North to South, had shifted from 40:60 to 20:80, and where the Boott had once run 151,000 spindles they now had 85,000 and ran only 75 percent of those.[54] A major contrast lay between the fiscal health of the corporation, on the one hand, and its prospects for continued manufacture in its buildings, on the other.

The unionization of the Boott introduced a new element to the situation. It enabled workers to act in unison, to make their opinions and demands known, and better to defend themselves against encroachment on their rights or their concepts of reasonable expectations. Yet wages continued to rise and fall according to factors outside the mills, and the union brought labor and management together in the MLC where they could jointly endeavor to improve the operations and its efficiency. In several instances, the union even strove to impose management's position on reluctant workers.

Little changed. The Flathers continued their intense efforts to pursue markets and to achieve maximum production from the plant and employees; in effect, they combined the external pursuit of niche-markets with the internal techniques of mass production. The directors continued to limit the pursuit of success to these tactics, to avoid investing in improved conditions or equipment. For the workers, wages remained low, the work both unattractive and frustrating, and jobs elsewhere (or even unemployment) preferable. Conditions at the Boott hampered their efforts to apply their abilities to the work. Their position relative to workers in other industries continued to decline.

Part Five

The Boott Mill Expires, 1947–1955

Chapter Eleven

Manufacturing Winds Down

IN 1947 the war was over, but the conditions it had produced remained at issue. Relations between labor and management now carried not only the weight of the operation's long history, but also new pressures: conflict born of industry's high profits during the war while wages rose slowly; increasing southern competition in a widening array of goods; and national drives to pursue labor's interests.

Products

At the Boott corduroy re-emerged as a major product, now finished at the mill; most other producers sent their material out for finishing. Its production rose to 30,000 yards per week, then to 60,000 by 1949. Dyed, high-quality goods offered a successful product, leading Flather in 1950 to cite "two excellent previous years" and to claim the largest corduroy garment-cutter in the country as "theirs." As prices rose, they added finer pinwales and printed cords for women's wear and heavier carded (not combed cotton) cords for men. At the same time, this involvement with fashion and consumer goods incurred risk; one of their customers began sending his car to the mill daily to buy yardage for orders he had received, unwilling to stick his neck out in a weakened economy.[1] Taking the bitter with the sweet, they enjoyed substantial sales of finished goods bringing top prices, but now operated in an area of the economy where they had to ride market swings along with customers directly involved at the retail level.

Toweling yardage, which had fallen off sharply when the mill had the opportunity to profit from the higher-priced military materials during the war, resumed its earlier ten-million yards a year level. Able at times to underbid southern mills for government contracts, more of their long-wearing towels went to institutional users or suppliers (62 percent), with retail (24 percent) and wholesale (13 percent) buyers accounting for most of the rest. Chainstores accounted for one-third of toweling sales by 1952, following the highest level of sales overall in the mill's history. So great was this success that when the opportunity came to compete for more Navy twill, they submitted a high and heavily qualified bid in order not to interfere with their standard production unless exceptional profits were available. Even downturns appeared to be temporary, with a loss in early 1953 overcome later in the year.[2]

Three solid lines with limited variation of styles or products within each offered the basis for steady and simplified operation. A problem was the susceptibility of these products to general economic downturn: "Our lines go to retailers, wholesalers, towel supply houses, cutters and garment manufacturers, who are affected by consumers' cautious buying." But this problem resulted from the mill's success in a long-time tactic, developing lines to be sold (1) finished and (2) under the Boott brand name. Yarn-dyeing at the mill also served flexibility, as well as proving both quick and cheap.[3]

Although a "variety mill" before the war, the Boott now operated flexibly but on few fabrics. As they described themselves: "a compact unit like the Boott, with close contacts with customers, and flexible facility handling frequent small orders quickly, with close control on labor costs, enables us to fare better under present conditions, than some of the large less flexible outfits."[4] An order mill, but not a variety mill. The market for its operations was also promising. A Boston garment manufacturer used Boott corduroy in supplying Jordan Marsh, Filene's, R. H. White, Touraine Shops, Gilchrist, and R. H. Stearns, while a Baltimore maker supplied Montgomery-Ward. Boott curtains were sold to supply houses and to department stores, such as Macy's, and directly, as to the Savoy Hotel in New York. Towels also went to wholesalers, but many sold through chains and mail-order houses. W. T. Grant, H. L. Green, Montgomery-Ward, G. C. Murphy, Sears, Roebuck, McLellan, Newberry, Enterprise, Kress, and Woolworth's (63,000 yards per week) all offered Boott towels.[5]

OPERATION

After a year under the new (1946) contract, management boasted that despite the 15-percent wage increase, negotiated contract alterations permitted increased production and actually *lowered* labor costs by 15 percent in the grey mill, which included half of their employees (down to 898, total) and payroll. All other positions ("finishing corduroy, scrim, toweling, stitching, maintenance, clerks, supervision, warehouse and power") awaited study for comparable changes. The pay increase, on the other hand, along with curtailment in war work elsewhere, led to more applicants for work and permitted more selective hiring. Thus it improved the quality of their workforce with a limited effect on production costs, much as John Rogers had predicted.[6]

Despite this success, management demanded workload increases of up to 700 percent in a sudden effort to attain the Fall River loads for existing pay. The workers voted unanimously to strike and initiated what turned out to be a seventeen-week conflict. Doffers and spinners had gone to arbitration months before rather than accept new loads, and the union insisted on Fall River–New Bedford wages without changes in loads, claiming that the workers performed to the level permitted by conditions.[7] Management wanted to be able to install "scientifically set work assignments" at any time, leaving the union only the right to grieve them while workers attempted to fulfill them. The union wanted to be able to go to arbitration before changes were instituted. It also wanted seniority to rule, while management wanted to use merit equally as the basis of determining employee preference. Management felt that shop stewards resisted changes and wanted cooperation. It sought Saturday operation without overtime in order "to use our waterpower economically." The Textile Workers Union of America (TWUA) wanted a union shop, five paid holidays, increased insurance and vacation pay, and a 5-cent-per-hour advance. It also asked for a clause to sidestep the new Taft-Hartley strike liability, a clause common to the industry in the area.

Management's objections were essentially twofold: it wanted any pay raise to be strictly tied to increased production; it felt that for the union to have the right to approve assignments in advance would be to hand over to it the management of the plant. Any advance in pay not tied to an increase in production, or any pay attached to an abstraction such as a holiday, it continued to consider "paying for work not performed." All payments, it believed, should come in the form of wages, not deplored "fringe benefits," and should come only when the work was stretched out or speeded up, or both. The directors were comfortable with this position, spending the strike sitting out a difficult time in the market and looking to make the operation more efficient at the same time. As long as they could avoid excess bitterness on the part of the workers, they were in no hurry to settle. They believed their willingness to pay the raise asked, with new loads, should prevent ill-feeling. This willingness did not, of course, put food on the strikers' tables. Workers were forced to depend on others in their families, to commute elsewhere to work, and to accept nearly any available employment.[8]

No real progress occurred until March 26, 1948, after management began a series of regular meetings with the local union head, Hodgman, and Herb Payne, a CIO vice president and top negotiator for the East. Payne wanted to work out a settlement on loads that would permit reopening. The Boott had strongly opposed the union shop, with Flather even sending a letter to employees before the vote urging them to side with the mill against the union. However, their "open gate" policy had brought few workers back to the mill in the months of the strike, the union shop vote (required by Taft-Hartley) at other mills indicated the futility of opposing it, and the time to settle arrived. The contract was approved April 17, 1948. The two-year contract called for rates of pay based on the Fall River–New Bedford peg-points, plus a 5-cent-per-hour raise, plus 10 percent, with a 97-cent minimum. It included a union shop, a nonliability "no strike" clause, dues check-off, improved insurance, vacation, and holiday benefits. Work assignments would be established by a jointly selected engineer during the first year, by the company in the second year. A three-man tribunal (one from the union, one from the company, and one chosen by these two) would hear grievances, but assignments would be exempt for the first year. Seniority would determine who would work, with those who had crossed the picket line included in the list on the same basis as the rest, not given priority. The owners saw freedom to increase loads as the primary issue. It promised competitive improvement and increased profits. The 1947–48 strike had demonstrated the extent of this belief when the treasurer and directors agreed to hold out for specific workloads in the contract while willing to grant most of the rest of the items at issue. Without such contractual specificity, they had feared, the union could "keep something boiling all the time."[9]

The contract granted management the rights it had sought: to set loads with three-day notice to the union and to work the assignments until arbitration rejected them, a freedom in excess of that obtained by other mills. By "strengthening their hand" in this manner they gained the ability they had long sought to evaluate jobs and set loads on a continuing basis. When operation resumed, the workers came back to "full loads" without difficulty, even among stewards who had led the opposition. As Director Bemis pointed out, it had been "very wise not to show we were in hurry to end strike." Another director stated, "My guess is we're ahead of the game." In each major disputed category, results matched their hopes: fourteen spinner-doffers worked the full assignments, with six over standard; three of four twister-doffers exceeded the standard (10 percent above the old standard), "including McDonald who testified that she could never do more than previously"; five loomfixers worked the new loads, representing increases of from sixty-four to eighty-four, from seventy-two to eighty-four, or from fifty-six to seventy-three looms, depending on circumstances; seventeen weavers tended the new standards, with two producing cloth above the standard. Helpers had also been eliminated.[10] Of course the small numbers of people involved reflected the early stages of the start-up, with senior people recalled. Also, the performance did not reflect the six of ten able to exceed the standard that the union had previously insisted on.

Management recognized that pay raises led to improvement in labor's effort and attitude. It detailed the effects of wage and productivity increases in a "Confidential Supplement" to the *Directors' Report* for May 1948. The Flathers indicated that to offset the pay increases (wages and fringes) in the 1948 contract required production of 78 pounds of cloth per week per person, up from the 68 pounds per week of July 1947. By June 19, 1948, they "reached 89 lbs. per man, per week!" In other words, they had significantly overcome the effects of the pay increase and turned it, and its accompanying loads, to their advantage. Put another way (which they did), if 1947 were assigned an arbitrary index of 126 to represent production per person at a given rate of pay, an acceptance of the CIO's demand for a 15-percent pay raise with the old loads would have equaled 145; paying the 15 percent with loads that "we interpret equal a 20 percent reduction in cost" produced a figure of 100; in practice they found that if July 1947 = 125, then April 1948 = 126, and June 1948 = 109, a decisive success. A year later they reported: "We believe our higher work-loads, better employee efficiency, and smaller amount of indirect labor perhaps result in a lower cost at Boott than could be expected in other New England mills." In other words, they were doing better with the union than they had without. By August of that year, after a 31-percent *increase* in payroll, there had been a 10-percent *decrease* in costs per pound. A 4-percent increase in workers had produced a 46-percent increase in pounds produced. Later that year, improvements in loads in finishing again increased production faster than wages. Between 1947 and 1949, despite feeling that they had raised wages too readily, without sufficient increases in loads, during the time of a 100-percent increase in the price of goods, they reported they received substantially more product for wages:

	5/47	6/49
Units/hour	73.4	99.7
Earnings & pay/hour	$1.0004	$1.178
Avg. cost/100 units	$1.368	$1.184

Increased production per person far outstripped pay raises and reduced the cost of making cloth. Similarly, production from hourly paid jobs rose 125 percent between 1948 and 1949. Savings in production labor costs occurred steadily through at least 1952, and even when wages could not be cut, as desired, savings appeared available from "improved assignments and rearrangements of groupings for combined supervision."[11]

They had created a new and more precise, and potentially more pressure-filled, technique for utilizing supervisors to enforce performance, more efficient even than that of paying piecework to supervisors: "We have a budget control check for employe's product. Each supervisor keeps informed of employe's production, and all supervisors are endeavoring to keep within their budget." Instead of receiving bonuses for exceeding goals by rooms, supervisors monitored individuals' ability to fulfill expectations. In a harsh extension of the old practice of piece-rates for overseers, in the mill's last era, raises for salaried employees were to be achieved by drawing upon money saved from declines in the labor cost of hourly and piecework labor. The ultimate guarantee of the exploitation of the latter by the former appeared to have been achieved.[12] They found that within all the new conditions and relationships, a "healthy and respective [*sic*] attitude" was found among both employees and supervisors.[13] As one Boott worker remembered, however:

> My mother never spoke ill of anybody that she worked with except their supervisor. . . . They got to know him so well that they swore at him. He had a hard job because he had the owners behind him telling him, the output's got to be this much, and you work those people until you get it! So he became the bad guy. Even if he wanted to be a good guy, he couldn't; it meant his job.[14]

Despite its successes, management bewailed the effect of the union. The strike in 1947–48 "hurt us and weakened this mill, and is in no small way responsible for our present crisis," it claimed, inexplicably, despite the "splendid job" the company had done "streamlining the plant." Not only was the CIO the scapegoat, but the blood of the goat washed away from memory the disastrous effects of decades of operation without significant repair or replacement of plant or machinery. Management objected that the union brought up inefficiencies cited by a production engineer, "rather than helping improve performance." Workers were at fault, for "the Northern textile operator through habit, tradition, and the teachings of the old craft union and his father before him" thought of the number of machines assigned him, not the time and effort involved. Yet the Boott operated the machines used by the workers' parents, expanded loads, and still objected to difficulties stemming from the bogeypeople of craft unions. Flather had written in 1936 to Cyrus H. McCormick, his old employer, that "as the whole industry is able to produce more than the demand, there is only a small margin for profit." Overcapacity, created by those he called "adventurers," caused the problems. But by 1951, a new source had been identified: "As a result of labor troubles and high costs, resulting in a great part from increasingly high wages and fringe benefits, the number of people we can employ even when business is booming becomes smaller and smaller." Wages, not the dozens of workers repairing the buildings, not the outmoded machinery, the small rooms, the divided machine assignments, the industry's overcapacity, none of the things identified by the Boott management and its consultants, nothing but the CIO was to blame.[15]

Despite the high ratio of profit increase to wage increase during the war years, management immediately considered those profits a thing of the past with no need to offer raises if current conditions did not "justify" them through increased exertion from the workers. Since Depression lows, Boott wages had increased 100 percent, management claimed, while the cost of living had risen only 90 percent, allowing labor an actual benefit of 10 percent over the desperate poverty of the 1930s. Others measured slightly larger increases in the cost of living and also pointed out that United Auto Workers members had *tripled* their Depression low by 1948, with half of them owning homes at that time, a situation far removed from that in Lowell. At the Boott, wages remained low while profits and salaries soared.[16]

When the cost of living plateaued temporarily in 1948–49, the CIO lost its case for a Fall River–New Bedford raise in arbitration and withdrew a parallel suit against the Boott. Later in 1949, although management saw an opportunity for a pay raise, the "CIO is not asking," so the opportunity passed. At the end of 1949, management paid each worker a $12.50 bonus because "we felt like giving a bonus to everyone."[17]

Early in 1950, prospects appeared bright. The Net Quick was rising toward a goal of $2,000,000. Stock was high; Flather said he had recently paid $142.50 per share. Five percent dividends reassured the directors, who described the Boott as a "pretty good enterprise," and said, "I'm not in favor of selling," and even, "Wouldn't know what to do with the money. Rather have any money in the Boott." The only question was whether to pay out 70 percent of earnings as dividends, or only 60 percent, the average for Dow Jones companies. Times appeared flush, and the paying of a fifth dividend for the year was "justified."[18] The CIO sought a 15-percent raise in the summer of 1950, citing the "action of Chrysler" and the 8-cent raise given in the South. This instance revealed a new response to such requests, with one mill agreeing and "forcing" the others to follow. In this case it was Royal Little, recent creator of the Textron conglomerate, who gave in first. Given the wartime (Korean) conditions, the Boott did not mind the 10 percent granted and noted it would help them retain workers. However, they felt betrayed by Little and others: "Now that labor, politicians, and employees have side-stepped their duty of using their influence to hold down selling prices, wages, cotton, and other costs, there is no restraining influence and we are on a wild ride." All factors were expected to assist the Boott, as if what was good for the Boott was good for America.[19] These groups had neither reason nor ability to hold down costs to benefit owners. For the Flathers, however, the conduct of the state represented a betrayal, and it served as another scapegoat, particularly in the mill's final years.

Rather than deal with the fact that the Boott couldn't attract male employees (unhampered by legal limitations on hours), they claimed that the "six o'clock and eleven o'clock law prevents 3-shift operation in Massachusetts." The forces arrayed against them appeared conspiratorial: "Cooperation between the legislature's and the union leaders has placed Massachusetts out of line with competitors." Politics became the bugbear for the mill while it struggled against handicaps in every area of its operation, from plant to machinery to workloads: "The fact is that we have among politicians and a legislative policy which is out of control, as far as meeting competition is concerned, and it has been killing off the textile industry." Both consultants and internal evidence had long made it clear to the mill's owners and managers that their policies, not those of politicians, were not designed to perpetuate the mill. In fact, those policies, not politicians run amok, produced the legislation of which John Rogers complained. As Parker noted earlier, "The difficult relief situation is an important factor in the high tax levy."[20] Despite a concern for the employees which appears throughout his commentary, John Rogers's perception of his own interests as serving their good led him to desire to have their protection abandoned: "It is a wicket [*sic*] thing for this State to have a law on its books which temporarily satisfies the employees to the extent which the State should take the matter into their own hands to stop it and safeguard the employees' job." In other words, in order to lighten the burden Boott management had piled upon the mill's back, laws protecting workers from the effects of such conduct should be repealed in the interests of perpetuating the Boott, at the expense of the workers' well-being.[21] At the same time that management decried government's role, it ignored the increased comparability of wages, North and South (8 percent differential by 1951), and ignored the true source of its legislated burdens. Uneven employment, an industry tactic from the start, produced the need for unemployment insurance, just as workplace accidents led to workmen's compensation. Legislation defined community standards and taxes supported the result. Industry created the problems government treated. John Rogers objected to the requirement that industry bear these costs of its doing business. When a Boott Mill memo in 1950 cited high Workers' Compensation and Unemployment insurance charges as significant taxes on the operation, it carried the realization that a safe and steadily running mill would not have incurred such high costs.[22]

Management continued to take advantage of the freedom in setting loads established in 1948. After a wage increase in 1950, management introduced new weaving assignments with "no incidents." It again demonstrated its thinking on the relationship between the two: "More pay in our case implies more work." Corduroy weavers tending twenty-four looms in 1947 had returned to twenty-eight in 1948 and now received thirty-two as their standard. Careful time studies of loom-stops in advance of the introduction helped avoid problems. As had virtually always been the case for years, this stretch-out included no element of technological improvement.[23]

The year 1950 was described as twice as good as the one before, with income after taxes and depreciation of $167,871, including a $15,000 tax rebate. An extra dividend immediately assigned that cash to owner income. Yet by the end of the year cotton prices hit their highest level since the Civil War, nearly 50-cents a pound, and soon the reasons for optimism became less clear, the mill's actual position harder to ignore.[24]

In January 1951, the TWUA opposed the action of the newly formed Wage Stabilization Board (WSB) in limiting increases to 10 percent. Flather saw a plot: "When a wage formula was in the making it was said that the whole C.I.O. Union including auto workers (Reuther), and the steel workers (Murray), told the textile workers (Rieve), that they would back him in a strike if he led the fight against the wage freeze." While they felt that they were in a good position regarding prices going into the freeze, they opposed Rieve's demands, totaling, by their calculation, 55-cents per hour. While they felt that their unit of the union was learning to consider the mill's need to stay competitive, "autocratic" union leadership called for strikes at some New England mills. The result was a 7 1/2-percent wage increase and a few minor matters, but without the cost of living clause granted by General Motors and Textron. The Boott continued to achieve minor differentials from the standard contract, which they felt gave them advantages. The WSB converted the settlement into 6 1/2 percent plus a cost of living escalator, but the Boott refused to pay it because the union had failed to get a parallel increase from southern mills.

Hourly rates for the Boott at this time were as follows:

Doffer, Spinning Frame	1.41
Spinner, Ring Frame	1.32
Loom-fixers (Plain & Dobby)	1.745

Weavers, Dobby	1.47 Avg.
Weavers, Plain	—
Battery Hands	1.21
Truckers	1.165[25]

The range, from about $1.16 to $1.75, while above the minimum wage, offered low compensation to the employees of a faltering mill. The average, $1.38, fell slightly below that of other New England mills. At no time after the war did raises keep pace with increases in the cost of living. A year later, in the spring of 1952, arbitration cases at several other northern mills led to the withdrawal of the increase. According to John Rogers, even union people admitted they had made a tactical error by imposing the increase only on the North, but another element of the problem was expressed by Flather: "The trouble is there are always one or two mills who are temporarily enjoying a moment of prosperity and they give in . . ., ignoring longterm needs." In the case of Little, for example, "it forced him to close all his mills in the North."[26]

In a matter of months the mill was not buying "high priced cotton" or producing much cloth. Management cited uncertainty about the Office of Price Stabilization, low orders, and customer hopes for cheaper cotton as retarding the market and claimed it an achievement to have operated without a loss. They did not touch the Net Quick or the Treasury Certificates, nor pass dividends. When the "almost unprecedented difficulty in selling" continued, mills began to contemplate closing and labeled it "a time like '21." Without large debts or a big inventory of cloth, with a Net Quick higher than in 1951, and with cash on hand, dividends and optimism continued.[27]

In the first months of 1953, the comparison to early 1952 made things look good: the calm before the storm.[28] A director observed, "You still seem to be solvent, allright [*sic*]."[29] By the end of the year, the mill ran just two days a week, and suddenly "things are getting worse all the time." Northern and southern mills alike posted, or concealed, substantial declines, even losses. Trying to ride out what some called the "worst textile depression in history," some mills responded by running three shifts in an effort to break even. The Boott managed to sell at close to its list prices, and income gradually rose "much faster than costs," which had fallen off slightly.[30]

In a time of economic downturn in the industry, once again, the Boott management made its strongest push for accommodation on the part of the union to its "special situation" as a northern mill with southern competitors, a situation shared by all mills. They claimed, "No mill we know of has a clearer record of cooperation with labor than we, and never have we had a genuine cooperation in return." In actuality, they extracted that cooperation, as shown by their figures regarding steadily increased production per employee, which outdistanced even their expectations and reduced costs through the final years. Regarding their record of cooperation, the union found them "amiable, eager to cooperate," only when busy on three-shift operation. Otherwise, they were slow to pay agreed-upon wages, had "many grievances," were "most difficult to deal with." The union charged that the operation was designed to keep the plant, not the workers, going.[31]

FINAL DAYS

Like a cartoon character who runs some distance off a cliff before noticing it is unsupported, the Boott Mills found itself about to plunge into the abyss. In 1953, the cotton textile industry generally found itself in poor shape with a bad market and big inventories. Boott competitors, such as Cannon, cut prices to move goods. A major textile depression was taking shape, and suddenly all avenues for assistance had to be pursued. Most pursued was labor, the usual target, and much of what was sought were givebacks. As early as January 1953, Frederick Ayer was adamant about the need for increased loads, despite the lack of change between northern and southern practice in this regard: "What do you mean? We've got to do something about this. . . . If we can't do something, might as well close up shop." Others recognized that such changes would be counterproductive, that the company "must have an eye out to prevent drifting employes. In a word, a good portion of our relief should come from better efficiency" resulting from retaining workers from whom higher loads could be demanded. The tactic of increasing loads without improving conditions had been pushed as far as it could go. Recognition that other factors impeded production appeared in minor efforts to improve processes and layout, but labor remained the prime target.

Losses in mid-1953 were much less than in the previous year and provided encouragement. Hopes arose for assistance from Governor Herter in meeting competition, but nothing came of it, perhaps because he believed that being associated with a wage-cut would hurt him in an election year, as the Flathers assumed. Wages represented 40 percent of production costs; since management claimed it had minimized other costs, any improvement had to come from this element. In reality, labor's cost-share was increased by the necessity of maintaining the repairmen needed to keep the decrepit plant and machinery in operation and the supervision to oversee the many rooms. The lack of spending for other purposes, such as depreciation and replacement of plant and machinery, also increased labor's percentage cost.

Finally the dialogue shifted, and the operation entered a period of discussions that assumed the mill would close: "The employees and the community should be grateful to the stock-holders for staying with the industry [in the past] during bad years as well as good." Frederick Ayer began to object to the situation: "How much longer are we going to let this drift? It's just coming out of our clothes."[32]

While the union sought a new contract adhering to the Fall River–New Bedford pattern, the management asked for a 24-cent-per-hour giveback from wages and fringe benefits to lower its labor cost to southern levels. It protested the state's safety net for the unemployed: "It is a wicked thing for this State to have a law on its books which induces employees to openly tell us that the spread between the tax-free relief money on the one hand and what they earn on part-time work on the other, is so little, that they would just as soon loaf."[33] Not only did they advocate raises only from increased production, not only did they finally want a 24-cent giveback, but they also objected to an employee who received unemployment benefits, group insurance payments, or workmen's compensation, or who received holiday or vacation pay due them, for that would represent some people "being paid twice." Vergados suggested increasing loads through modernization, the technique of technological advance usually used to justify the stretch-out.

Management continued to see its relationship with the employees as a personal one, a situation in which it could best determine and define the workers' needs and best interests. It failed to learn the lesson of the industrial relations people they had consulted since the teens, that workers had interests which had to be considered and that mechanisms existed to facilitate the exchange, and that the exchanges were not personal, as they were felt to be by the management, but simply an effort to redress the imbalance of power that had been part of the Lowell situation for more than 100 years.

When the end of the mill was near, special pleading became stronger than ever before. In letters to the workers, management warned that work elsewhere wouldn't be as "comfortable or as home like," as at the Boott, and that closing bore dire consequences: "Some of you will have to just stop earning any money. Some of you are too old to learn a new trade. Some of you would never be happy anywhere else." They claimed concern for the workers' position: "We want to impress you at the outset that we sincerely *want* to keep operating *right here* in Lowell. We are local people—this mill and this city mean something to us and so does the welfare of our fellow workers."

In 1954, the union found proposed loads "beyond the realm of possibility" and would not consider them without guaranteed earnings. In fact, M. Michael Botelho, Assistant to the Regional Director, TWUA, stated the union's belief that its willingness to attempt the operation of the plant "under a maxima workload" with guaranteed earnings represented an offer any other company in the country would readily agree to and be grateful for. He continued:

> Many times in the past Unions have been indicted by the press and others ignorant in the matter of collective bargaining, as the responsible agents for the liquidation of plants where they may have collective bargaining relations. I am quite certain that in the case of the Boott Mill, should you and the Board of Directors elect to liquidate, the responsibility will rest solely with the Officers of the Company.[34]

A comparative study of the significance of successful labor organization in various industries, times, and places would clearly reveal the relatively minor impact of unions on the situation in Lowell.

Neither state nor federal conciliation services could affect the impasse. Labor doubted the owners' intent to close without the giveback, citing the Flathers' sentimental involvement in the mill, but also held that if the closing were inevitable, it might as well happen while jobs were available elsewhere. The textile industry as developed in Lowell did not offer workers an advancement path, did not inspire confidence nor provide satisfaction.[35] It did not offer the workers any of the advantages available to managers and owners. Both the Flathers and the directors were ready to "go to the mat" on the issue of the reduction, the separation of the mill from the New England standard. As Bemis put it, "Too bad union feels the way they do. No need to go on dropping money year after year. Either they see it or we have to do something about it." John Rogers responded, "If union refuses to allow us to compete, we'll change our form of business." Bemis: "That would be doing something about it."[36] The union resisted the loss, the precedent, and the potential impact on efforts to organize other mills, such as the Merrimack, where they were again mounting a campaign. The union doubted the company's statements regarding losses, but also felt that it should be willing to lose money for a time, given the profits it had made at others. Operations continued on a day-to-day basis, with talks continuing intermittently. In January 1954, they made the decision to discontinue Carding, Spinning, and Weaving and to use purchased weft to run out the warps in the looms. Carding would be essentially finished by May, winding, warping, and slashing by June, and weaving by September. The deed was done, the mill's fate, long since sealed, was about to be enacted.[37]

Even in the mill's darkest days in 1953, when the Flathers saw labor givebacks as the only hope for saving the mill, they could neither convince labor nor even communicate successfully with them. "We could, with Vergados' unqualified cooperation and employes' unqualified cooperation produce more with the same number of employes and go a long way towards balancing the budget." Yet they didn't say this to Vergados; rather they denied it to him until December, when they sought his assistance despite their history and despite their admission that "some" workers would not be able to earn the wages tied to the desired loads because of "inability, unwillingness, or *lay-out* [emphasis added]" of the factory. In other words, they belatedly sought aid from the workers in solving problems created by management in a plant which would prevent workers from achieving the inadequate wages offered. A more complete separation, alienation, of the two sides can hardly be imagined.[38] Discharge notices were sent to workers to avoid retirement/discharge pay for the few (up to thirty-four) who might have qualified. In retrospect it seems clear that these negotiations were doomed from the start, based as they were on the union's desire to protect the workers and management's intent to extract significantly more production from the mill despite its refusal to invest in the equipment needed to generate maximum production from minimum labor.

The result:

Mr. John Smith
45 Smith Street
Lowell, Mass.

Dear Mr. Smith:

We regret to inform you that your department has been permanently discontinued with the result that your job here is no longer available. We enclose, accordingly, a Suspension-Discharge Notice to this effect.

We are sorry that it has not proven possible to save your job, but as there is no longer any hope of doing so, we feel that it is only fair to so inform you, in order that you may secure other employment. We hope you will be successful in doing so.

Very truly yours,
BOOTT MILLS.
Personnel Department[39]

Assets were listed for the directors:

1. $600,000 in cotton and goods
2. Ownership in Locks and Canals
3. " in LIDC
4. Machinery

5. Net Quick
6. Carry-forward tax losses of $271,000 for 1952 and $311,000 for 1953; 12,500 shares @ $46.55
7. Markets
8. National trademarks
9. Plant
10. Power
11. Management

Several of these would be of value only if operations were to continue, of course. For a time, the directors waited. As Frederick Ayer put it, they were "willing to lose as long as the Flathers could afford to." But by May, he and Bemis announced the end of their patience and directed an end to the losses; it was time to cut it off. Alabama Mills, seeking finishing capability, made a careful inspection of the plant and equipment before deciding against purchasing it. While $300 per share ($3,750,000) had been mentioned as a price for the entire package, sale occurred in October 1954, for $130 per share ($1,650,000). After fifty years of operation as Boott Mills, following seventy as Boott Cotton Mills, the operation had run its course.[40] Several months later the buildings stood silent.

CHAPTER TWELVE

Two Points of View

MANAGEMENT'S ASSESSMENT

IN ORDER to assess the meaning of the Boott's demise, we must look back over the managers' descriptions of the events, decisions, and practices of past decades. On the one hand, they sought new sources of income, as when they introduced finishing in the thirties, turning an expense into a saving and, at times, an income from other users in this higher profit-margin area. On the other hand, they acknowledged on numerous occasions, that in addition to ignoring plant deterioration, they were not putting aside a depreciation reserve. In 1927–30, 1935, and 1939–40, seven out of fourteen years, no contribution was made to that account. By 1935, Flather noted that the mill was $2.5 million behind on depreciation, a position which "may not be so far off." Instead, in 1937 he reported, "We are thinking of slowly and gradually using some of it [the reserve] to meet some of our deficiencies." Given their conservative behavior in the areas of technological advance and plant upgrading, such use of the account would have been more of a stopgap than an actual offensive against the inexorable forces of depreciation.[1] It should not be assumed that Flather endorsed this approach altogether. In 1936 he was considering seeking Dutch capital to create a group of integrated industries, not necessarily including the Boott Mills, "which have suffered through the lack of an aggressive, progressive leading interest among the owners, many of whom are trustees of estates of former large and small owners."[2]

The conduct of the business took place within the constraints of cautious directors with extensive interests elsewhere; the manner of operation also appeared in accounting practices of deception and self-service. Statements were devised to hide profits and confuse those who might be interested, whether labor or government. Profits from changes in cotton values, a form of speculation, did not appear, nor did those from holdings in Lowell Industrial Development Corporation (LIDC) or Locks and Canals. In fact, profits, particularly during the Depression, were always calculated "after writing down inventory" of cotton and/or cloth on hand, an opportunity of delightful freedom. Depreciation always offset income, but, as noted, reserves for its effects were not given similar attention. Improvements, paid for by borrowing, were routinely capitalized, thus giving the owners profit from production and diminishing the apparent percentage income.

None of the above procedures requires ex post facto attribution of intent or practice. As stated in directors' meetings, Reserves for Contingencies, Suspense, and Profit and Loss were various names for funds "washed out" before state reports were filed. At times, these represented large amounts of money in proportion to acknowledged earnings, as in 1938, a year with a reported $6,000 loss after $52,000 in flood damage repairs had been subtracted from earnings, when Director Roberts observed: "Your Reserve for Contingencies is large. This $400,000 is really profit," with which Flather agreed. Along the same lines, they regularly used reserves for unearned dividends, which themselves were manipulated to disguise the true return on investment. Flather outlined the situation in describing "Depreciation Reserve and How It Got Started":

> We paid $300,000 for plant in 1905. Was set up on book at 600,000. That gave them capital surplus to go on. . . . Since then every dollar spent on plant has been added & capitalised. The old crowd couldn't get out to [sic] fast before it failed. So they closed in 1904. Probably plant was on old company at some 3,000,000. So new company was justified in marking up 300,000 to 600,000. It wasn't "water." If same amount of work & effort had been put in down South it would have earned more. . . . Some years we couldn't afford to set up depreciation.

They could nearly always pay dividends, of course. In any case, as the Boott generated money, that borrowed for the equipment was treated as if it were from the investors. The difference between three and six hundred thousand dollars certainly was water, or fictitious investment, and so was the rest of the later $1,250,000 self-valuation which was used as if it represented investment. The reported returns actually concealed a much higher rate of return, perhaps four times as high. Furthermore, the high figure was said to represent a realistic appraisal of the value of the shares involved, $100 each. When the shares traded for as little as $33 in the thirties, dividends began to be reported in dollar amounts which, while they equaled the payments made on a percentage basis, would have appeared as triple the size if reported as a percentage of the shares' market value, or as much as twelve times their reported return on investment. As usual, real income was concealed from all but the directors.[3]

In addition to squirreling away reserves to hide income and manipulating figures to diminish its apparent dimensions, the Boott was also used as a money cow, a source of funds to be directed at other ventures rather than toward its own perpetuation. As Flather said in 1951, "We have worked along the line of conserving cash, which leaves us in a position to go extensively into anything profitable that may come up in time. On this principle we have quite a few times in our history turned our resources into something profitable." The profits derived from other investments resulting from this policy did not appear in company reports and the enterprises can only occasionally be identified. They built a service station on their land in the early thirties and leased it; they united with the Merrimack (controlled by the same investors) in the LIDC to buy up the real estate of the Massachusetts Mills, along with its waterpower rights plus those of the Lowell and Middlesex Mills and the 622 shares of Locks and Canals associated with them. These moves not only saved them large sums on their expenditures for power, but the Boott also wrote off paper losses on these investments against profits without including income from them on their balance sheets. Neither did they include their profitable and growing share of Locks and Canals holdings. In fact, after writing off losses as high as $100,000 per year, the Boott moved to sell the LIDC to Locks and Canals when it reached a point in 1952 of producing a net profit of nearly $31,000. The new money cow needed to be hidden in another stall.[4]

A description of the treasurer's role at the Merrimack during Flather's tenure there offers insight into the nature of his administration:

> The principal defect in the organization is one common to small organizations headed by an unusually capable and forceful person, namely, a concentration of responsibility in the chief executive which tends to discourage initiative among subordinate members of the organization and in turn to induce a lack of confidence on the part of the executive in the capability of the subordinate persons.
>
> It seems to us that the treasurer has lacked confidence in the manufacturing management at Lowell and in the representatives of the sales agency in charge of the Lowell product, and has undertaken, alone, responsibility which normally would have been shared by an organization in whose abilities he had complete confidence.

It can, on occasion, be seen at the Boott that "the boys," as he continued to call them, hesitated to make decisions in Flather's absence or without his approval. While he gave them credit in meetings, they did not appear independent. Instead, it seems that they may have attempted to make up for lack of real power with a meticulousness which mimicked their father without infringing on his prerogatives. In 1945, after serving as plant manager in charge of manufacturing for over ten years,

Frederick continued to involve himself directly in matters concerning starch, lights, building conditions, loom details, spare parts, inventions, packaging boxes, toweling promotion, bobbins, shuttles, sale of second-hand machinery, thread, paint, oil, etc. This level of detail was very time-consuming, much of it belonging to a purchasing agent and involving matters of a few dollars. As sales manager, John Rogers served as liaison between production and distribution of the mill's products, but he also oversaw personnel work and labor relations. Applied to issues of safety, their attentiveness improved the mill's record substantially. In other areas, their attentiveness became distracting.[5]

Dealings with employees did not proceed on a basis of honesty from management. When the WLB required a wage increase for mills paying below the area standard, the Boott attempted to turn the implied reprimand into an example of largess and claimed credit for a "voluntary" pay hike. In another instance, union officer Bishop discovered that they had inserted changes in a contract supposedly simply being reviewed, and "We gradually had to relinquish our position on several points we were anxious to incorporate," because of being detected in the subterfuge. While aware of the extent of the employees' dependence on their income from the mill, management noted that it hadn't occurred to Flather to spread out the work, to curtail partially rather than to lay off employees. While Frederick had once claimed such thoughtfulness, in 1931, before he had gained sufficient power to make it happen ("Our prices are too low to make any profit, but it does provide employment in Lowell"), at about the same time John Rogers had responded to a suggestion that their running was "welfare work" by saying that things were not quite that bad. In general, they were quick to go to short or no time in parallel with the condition of the market, and when they added a little production for less immediate sale than usual, it was to "keep employees on and save money from losing and training" help.[6]

The ability to operate the factory in this manner stemmed from three sources: the perceived nature of the economy, the workers, and the city. Turning to the latter two, for the moment, we discover that management held opinions which included the belief that capitulation to loom-fixer demands would "add to the use of liquor," that grievances were likely to represent "showing off for the MLC," and that absenteeism was still considerable (but "it is probably true that a large amount of it has been caused by sickness following a severe winter"). They believed that there was "a disposition on the part of some to loaf occasionally because they feel that after they have reached a point in their earnings, part of their money will go out in taxes. Tax dodges were for investors, not workers. The most fully stated and harshest view appeared as the preface to an article by Frederick, "Maintaining Safety

Programs During War Production," in the *American Wool and Cotton Reporter,* a title which suggested concern for worker safety:

> With the average mill operative receiving perhaps two or three times his former average earnings and at the same time working comparatively few hours for his money, which is far in excess of any rise in living costs, he and she have a strong urge to spend a large part, if not all, of their manna from heaven. Perhaps much of this is understandable in view of the many lean years textile operatives have been through. Nevertheless, the spending on excessive food, drink and entertainment has resulted in marked increases in absenteeism, digestive disorders, physical fatigue and mental depression.

In his view, not only were wages "manna from heaven" rather than earned income, but the nature of the employee was to misuse the largess to his own, and his employer's, detriment. Such thinking led to the mill's place in Lowell being viewed in parallel fashion:

> Boott Mill is ideal for Lowell people. . . . The bulk of Boott employees are help not adjustable to other industries. This is evidenced by the fact that 86 percent of Boott Employees have been at Boott over one year, not withstanding the fact that other concerns, offering various inducements, have failed to interest them.

While in one breath only recent high school graduates objected to conditions at the Boott, in the next anyone who worked there represented proof of mental or physical impairment, a special charge on the philanthropic nature of the corporation, without which the city would be left bereft of an employer of last resort.[7]

In the mid forties John Rogers made statements to the Tax Board which he knew to be less than frank: "The city needs us. It would be an awful situation if we should move. We have no intention of moving if we can possibly avoid it but we need all the help we can get from the city and from the state because our buildings are a handicap." The statements ignored the severely limited nature of the Boott's commitment to Lowell operation and sought help under false pretenses. Demanding a 24-cent-per-hour pay cut, management told the workers to fight their union: "This is *your* mill. It is *your* job that is at stake." Repeated letters from the Flathers to the employees denied the reality of the situation and obscured the identity of the interests which were being served.[8]

The importance of management style in employee relations did not escape its attention. When the Navy told management to inform the workers to eliminate faults "or else," they replied they couldn't, that "we had to treat people in a way so they would be willing to work here rather than go to higher wage plants." They wished to substitute style for pay. Flather endorsed Royal Little's condemnation of the young Dumaine at Amoskeag when he described him as worse than his father, "like his father in roughness and walks up and down the office swearing at everybody," an improper manner in which to treat those from whom one wanted more than one would pay for. The position of Employee Counselor was created to treat workers in a manner comparable to that in which a nurse treats an injury; if a worker became upset, he or she would be sent by the supervisor for counseling related to "suffering a mental disturbance," the implication being that grounds for such a feeling were unreal. Employees were to be protected from outsiders, such as immigration officials, whose questionnaires worried "highly esteemed" workers: "It is not right for us to worry innocent people." They were to be befriended: "We discussed the history of the French people in Lowell, and how much we liked them," with a Franco supervisor and union man. They were to be reminded of their role in the national defense effort, when all workers received pins displaying the "E" for excellence awarded the mill by the Army and Navy for the Boott's conversion to war work. Such treatment fostered an illusion of joint effort, despite soaring profits and restricted wages. More substantively, John Rogers, a member of the draft board, abstained from voting while two union organizers were drafted. Both were exempt, one as an active labor leader, the other as the supporter of his mother and three younger brothers. John Rogers described a likely rationale: "Mr. Vergados left saying that War is a slaughter of the masses by a few big men who desire one form of Government."[9] Active union leadership and offensive political beliefs outweighed the two men's legal rights.

Inherent in all these comments and actions was a belief that labor was not a partner in the operation, that the interests and decisions of the managers and the investors were independent of and superior to those of the employees. It permitted the decision to allow the mill to gradually deteriorate over a period of time while the equity the employees had helped create was used up for the exclusive benefit of the owners.

Satisfaction with and endorsement of the Flathers appeared consistently in company reports. Confidence in Flather, along with the incestuous nature of the industry at Lowell, appeared in the move by the directors of the Boott and Merrimack (one group of people alternating hats) to install him as treasurer of the latter mill, as well as the Boott, in hopes he could rescue it from the losses which had it headed toward bankruptcy. Obviously they thought well of Flather and lacked the hindsight opportunity to assess his ability to evaluate the situation in 1939, when he attested that southern mills were losing ground, that Roosevelt could not be renominated, and that war was unlikely.[10]

Yet John Rogers described as "tragic" the three results, particularly for the middle-aged or older workers, of the Boott's closing: "1. His lifetime skill and experience no longer useful. Having to learn a new trade physically upsetting. 2. He is upset mentally to have to start on a new job, if one can be found. 3. He is terribly upset financially."[11] Despite such disastrous effects, the company need not operate as a self-perpetuating entity.

Labor's Interpretation

The distance between the point of view of management and that of labor could scarcely have been greater. As Irving Bernstein so aptly defined the situation, the textile workers' situation was one of constant fear. With good reason they feared the plant's closing, wage reductions, workload increases, discharge for union activity, the effects of cold, inadequate income for food and clothing, poor health interrupting poor wages, and an uncertain future. The owners and managers shared none of these concerns.[12]

J. William Belanger, head of the TWUA for New England, offered a formula for success and an analysis of failure:

a. Modernize plants
b. Plow back profits
c. Develop new fabrics

He characterized the New England industry thus:

d. Manipulators and wreckers
e. No interest in continuing textile production in New England.
f. Squeezing money from the business.
g. Liquidation
h. Coolness of banks [to the industry here].[13]

While the Flathers denied the applicability of these charges to their operation, they fit the conduct of the Boott, both in the long and short term.

Much of the record of the workers' responses to conditions and events at the Boott can only be found in their action: striking, protesting, unionizing. However, recent interviews with Harry Dickenson and Vergados, the former a loom-fixer and both union officers, offer commentary from labor on a number of issues. Brody has noted that "labor felt grieviously short-changed during the war—hence the bitter strikes during the reconversion period." Vergados added a further element to the nature and extent of the resentment: "I read the financial reports, they were paying dividends when they said they were not making any money."[14] Vergados also indicated the place of a worker in the outlook of management: "They feel that you are not the human type, human beings that would stand up," a judgment of the situation in 1947 when workers were insisting on their right to a share of the surging income of the war years.

Dickenson's comments about entering the mill offered another view and an interesting counterpoint to the resentment of Agent Cumnock and his associates over the Commonwealth's belief in the existence of a permanent class of mill-workers:

> No. He didn't have to teach you, you were suppose to be in the mill. What was good for your mother and sisters, you were suppose to be. That's the worst part of it, I think that is the worst part of any family, you don't give the kids a chance to get interested in anything else. You have to be a mill rat. That's what they used to call them, mill rats. My brother was awful strong on that, he was all mill.[15]

Dickenson indicated the extent to which going into the mill represented thwarted ambition, the loss of an opportunity to choose, caused in part by the family's need for the income of every member to survive. The necessity of earning even low wages meant an early end to school and little opportunity to learn a trade. He also reflected the decades-long attitude toward textile work as a last resort, not an employment of choice.

Dickenson's account also reflected labor's concern for the operation of the mill, its knowledge and frustration, as well as the somewhat rose-colored memory which cast the best light possible on the work that had been his life:

> Some mills you know are run better than others on account of management. The Boott, we used to go to work there and I'd rather go to work there than go to the show. Everyone was pleasant. The head of the mill, he started putting his sons in there. . . . Before, they used to hire a man to run the mill that knew the business. Now, when they started putting the sons in there that had come out college it was a different story altogether. . . .
>
> What he [Frederick] did was breaking down the standards of foremanship. He'd take a man off a junk pile and put him in the mill as a foreman. He really ruined the mill. But John he was older than Frederick, you could talk to him, he had more reason. Frederick seemed to take charge and you couldn't do anything about it. When I was foreman, you know in the mill you have got to have heat, warm, and then you've got to have humidity and you can't have no drafts. The drafts in the weaveroom causes static electricity and it dries everything up. . . . So, I had trouble there and kept asking for more heat but all they did was just laugh at me. Some of the weavers were having a hard time and there was a place there, next to where the weaveroom was there was an empty room and it was wide open. Every so often there was a window there, it was open. So the weavers couldn't do

> nothing. You know what I mean, work hard for nothing. So I spoke to another guy, I was fixing looms then. I said "Give me the right to close those windows. . . ."
>
> He [Frederick] came through and saw it. Ordered them all down, all the windows open again. So, you couldn't do nothing.[16]

Dickenson's commentary reads like a summary of the many hired consultants. His evident satisfaction in his work when it went well, and his obvious frustration with poor operating conditions, paralleled the advice received for decades. Heat and humidity problems had long been noted, as had the unpleasant working conditions related to them. Questionable management practices appeared, but the investment policy of the Boott directors prevented any actions aimed at lasting success. All consultants noted the workers' disaffection, and the attitude and conduct of the employees displayed anger, contempt, and uncooperativeness. When the Navy rejected great quantities of uniform cloth for faults, Vaillancourt expressed agreement, said the cloth was "junk." When stewards were chosen, the Flathers felt they had been "selected from among the incompetent quarrelsome fringe of our mass of able and willing workers," an easier concept for them to accept than the fact that these people had been chosen by the workers to *represent* them. Had they been the happy workers the Flathers envisioned, an outside examiner would not have characterized the fixers as careless, willing only to fix, not to maintain, the looms, not giving full hours, wasting time, putting their dirty hands on the cloth, not even cooperating with one another across shifts. Willingness to work, to contribute, had been stifled by years of frustration and mistreatment. Labor saw the factory as a place of contending forces, not an enterprise to which they contributed their effort to achieve shared ends, mutual purposes.[17]

CONCLUSION

Aggressive pursuit of market niches such as towelling and curtains, and marketing to garment makers and retailers, along with wartime demands, had sustained the mill in its last decades. Although the die had been cast in the twenties, fate had been fended off or held at bay. Flather devotion and effort had done all it could. Finally the nature of the situation had had its inevitable effect. The workers, long martyred by the conditions of the buildings and equipment, could no longer bear increased burdens. As Loper had pointed out, not southern wages but Boott conditions put it at a disadvantage, and labor, finally, was unable to make up for them. The loom-fixers' rejection of the idea of competing with the South on the basis of living standards and the mill's inability to attract new workers marked the limits of Lowell's revolving-door labor supply.

An exchange of letters between Joseph J. Higgenbottom, a long-time office employee, and Flather in 1955 reveal two sides of the thinking about the Boott's closing. Higgenbottom's regretful letter blamed the union, previous to which management had been able to settle all labor problems: "We fell down in 1937 and again in 1939, only because the country . . . was being misguided by the group then in Washington. If our people, had not been led astray by outsiders, the Boott Mill would still be running."

Flather knew better. In an equally friendly and gracious letter, he recalled an association dating from before 1915, at which time, he remembered, Higgenbottom had been the "leading spirit" of the company baseball team. But of the closing, he wrote: "It was absolutely unavoidable and beyond the power of any human agency to prevent, but the staff have always been a bright element in the picture." In his analysis, the closing was "fifty years overdue," the product of the decisions and approaches I have been describing, not of unions or any other recent development. He had spent fifty years fending off the inevitable, and he knew it, and found comfort in the effort and the knowledge that nothing he might have done could have altered the outcome.[18]

According to textile editorialist Frank P. Bennett's memoirs, Flather brought merchandizing zeal and innovation to the Boott and effectually displaced the selling house from the operation. Other mills simply failed to make the effort: The mills in New England had become, by the time of World War I,

> "old—old buildings, old machinery, old methods, old managements, inherited jobs, the old commission house which was just a headquarters for the transfer of cloth (and the billing) without any real 'merchandizing' effort (but with a sufficiency of commissions to those 'houses.'" Mill Treasurers inheriting big salaried jobs; cotton brokers and wool dealers favored by those well paid Treasurers, thus getting that very profitable business with little or no competition and favored banks getting the mills' balances with the latter generally with no practical knowledge of the intricacies of textile operations and, finally, an almost utter lack of interest or any aggressiveness on the part of the stockholders (the proprietors) or the banks, for that matter, to change the management of those big properties, to change the merchandizing or to quicken the whole operation to make their shares, their ownership, income-producing and valuable.[19]

Flather's energy and leadership helped him achieve an estate valued at over a million dollars at his death in 1967. Even while it was claimed that the Boott was not making money, it paid dividends of $50,000 per year and a similar sum in Flather salaries. In fact, the Boott did not fail. It was a great success, holding on longer than all but one of the original Lowell mills, pumping out money countless amounts greater than had been invested in it. Finally it closed because, as a Flather had stated in 1945, "capital is ambulatory," and it found the walking pleasanter in the South. Labor was not invited along. After years of blame and years of increasing pressure, labor was abandoned, in no sense a part, let alone a partner, of the essential operation, the manipulation of money. The Boott represented the success of the Lowell plan, but the betrayal of its claim that private interests equaled the public good, a betrayal of that justification of the right to re-order and control workers' lives for corporate ends. Ultimately those ends were solely corporate, and the workers stood revealed as the wage-slaves they had feared becoming, victims of, not participants in, the system which had taken over their lives.[20]

History reveals that labor and management interacted with full awareness of their divergent purposes. Piecework and bonuses are examples of the attempt to extort effort from the unwilling. The two sides had, and recognized, divergent interests. A system which reserved its fruits exclusively for one group among its participants could not earn allegiance. Thus it fulfilled the fears of its founders, who sought to obscure the nature of their operation with amenities and public relations. Workers with alternatives will seek milder representations of industrial capitalism, just as corporations seek cheaper operating conditions.

While other industries, because more capital-intensive or more collectively organized, avoided the effects of competition longer, the current movement into a service economy arising in their wake reveals to us the pervasiveness of the principles and effects of Lowell's development. This country still awaits an economy which recognizes "public good" as its goal.

The tragic flaw which guaranteed the Boott's failure lay not in a human agent, but in the true main character of the development, capital. Boott founder Abbott Lawrence wrote in 1835 that "money holders . . . can transfer their persons and property to any given place or out of the country, having means always about them to do so." For him and for Flather, the meaning was the same: money can do as it will, is unfettered *by anything*. The final irony is that Flather indicated that the 1905 reorganization left the operation burdened by indebtedness and was the product of borrowed money, meaning that the difference between directors and workers was that one could borrow and one must work.

Postscript

TODAY, a visit to the Boott Cotton Mills, the labyrinthine complex constructed from 1835 to 1900, elicits thoughts of the tens of thousands who worked there during its 120-year history. It stirs feelings of profound sadness. These people did not have the opportunity, through their hard, efficient, and talented work, to contribute to the success of an operation that would advance their cause with its own. They could improve their lot and that of their kin only by enabling them to follow a different path, and they had limited opportunity even to do that, as Constable George E. McNeill in 1874, and worker Harry Dickenson in 1980, pointed out.

I feel sadness and frustration with this vision of the Yankee, Irish, French-Canadian, Greek, and other men and women toiling in a situation that steadily directed the money produced by their efforts to other pockets, other endeavors: the banks, railroads, and Massachusetts Hospital Life Insurance Company of the early days, the "likely opportunities," the Bemis Bag Company, the Lowell Industrial Development Company, other unseen enterprises, and the owners' yachts and travels in later times.

Boott owners hung on for the "market upturns that would enable even old equipment to be run at a profit," as Mark Reutter writes of steel in *Sparrows Point.* They utilized the Flathers' dedication and merchandising to continue production in a doomed operation. The closing of nearly all the other textile mills in Lowell in the thirties flooded the labor market just as immigration had in earlier years, renewing the supply of cheap help on which the industry had always relied. These industrialists were not in the business of making a product, cloth, but of making money. Dalzell makes this point for the early days of textiles as clearly as do David Bensman and Roberta Lynch for steel in *Rusted Dreams.*[1] Both Abbott Lawrence, in 1835, and Frederick A. Flather, in 1945, had affirmed the mobility of capital, its detachment from a given enterprise or location.

The Boott combined with long-lasting result the tactics and strategies Philip Scranton found divided between Philadelphia and Lowell. In both cities, "owners scrambled for survival, victimizing workers, hunting for niches, dumping maintenance and equipment updating overboard." In Lowell, the Flathers demonstrated a "dogged perseverance" comparable to that of a Philadelphia proprietor, thoroughly mixing "firm and individual identity," yet at the last moment, spurred by the owner, the Flathers performed the "corporate surgery" that Scranton associates with the exit of textiles from New England. The Boott's owners had found in the Flathers that Holy Grail of capitalist operation, managers totally devoted to the business's success, completely honest in behavior, yet governed by rational analysis of events and prospects. The Flathers displayed neither the faults of the inattentive or disinterested executives of Frank Bennett's New England model (nor of the *Rusted Dreams* pattern) nor that unreasoning persistence of the Philadelphia proprietors and their sons.[2]

Recent writers often criticize business for being devoted to the production of profits, not of goods. They identify tax policy, junk bonds, savings and loan rip-offs, stock and real estate trading, and hostile takeovers as actions producing wealth for a few at the cost of jobs in manufacturing plants. They see managers as having lost interest in manufacturing for its own sake, shifting their attention to short-term gains at the expense of long-term survival and growth. The story of the Boott Mills indicates the lack of distance between older practices and recent "plunderers." The Boott took every opportunity to profit from activities other than manufacture, to shift capital generated by this enterprise to other activities, other locations. The conduct of the Boott seems very modern, or makes that which is now called modern seem very old.

The Boott placed extreme reliance, and burdens, on the workers for the success of the operation. Cumnock emphasized the workers' contribution and significance, while incidentally indicating their lack of reward: "Our mechanics and overseers are mostly poor men, and most of the inventions of any value have come from them."[3] Consultants noted it took greater skill for workers and overseers at the deteriorating Boott to keep the plant in operation than at better-maintained factories. Directors placed the burden on the Flathers, who couldn't offer the production expertise of employees lower in the pecking order, but who gave effort and attention where they could (e.g., the office, advertising). Variety and quality of goods produced didn't come from technology, which only declined, but from labor's talent and dedication. This reliance on workers makes all the more stark the contrast between performance and result, input and reward. The directors' policy doomed the mill and the jobs, despite the workers, not, as has so often been argued, because of them. David Brody notes the power of a pervasive management philosophy in dominating labor.[4] The concept that labor produces runaway industry is belied by the Boott, the story of which underscores once again the power of that philosophy which places blame on those who perpetuated the operation, not those who closed it.

The industrial style of operation developed in Lowell not only placed capital and labor in opposition but also alienated the mills from the communities in which they were located. Local and state governments found that they had to monitor and restrict industry and to ameliorate its effects. They made rules regarding school attendance, limits on hours, wage payment, and safety; they levied taxes to support unemployment insurance, workmens' compensation, and such—all brought on by industry's practice of hiring children, injuring innumerable workers, and employing a labor force intermittently. Business, not labor, bore responsibility for these costs. Consultants found unsettling implications regarding attitudes toward worker safety and protection, the absence of the "working condition they are entitled to and will demand."[5] Labor was always a temporary, transient part of the Boott's profit equation. In addition to the poor working conditions, the slope of the playing field was revealed with particular clarity by the directors' willingness to pay dividends, but not wages, out of accumulated earnings.

Robert Reich notes that "the business culture's inability to understand and respond to the civic responsibilities attendant upon a large corporate size helped inspire" repeated waves of political opposition—from Populists to Progressives and beyond.[6] Yet American industry, while regulated, taxed, and criticized, still operates as if under no obligation to the interests of the local or national community.

Workers in textiles bore the brunt of deteriorating conditions, yet at the same time they perpetuated the operation through their knowledge, skill, and effort. The industry's exploitation of these workers led to increasing levels of government intervention. The accompanying costs could only be evaded by continually seeking an arena in which this overhead had not yet accumulated or where it would be ignored.

From the beginning, owners at the Boott turned away from styles of production which maximized the contributions of employees' mental and physical skills. They pursued a line of development that spread labor thinly but intensively over increasing numbers of machines. Operating the complement of equipment representing one mill at the Boott required 226 employees in 1838, 90 in 1876, and 39 in 1910; the decrease related far more to what could be demanded of workers than to labor-saving alterations to equipment.[7] Ownership avoided, even prevented, workers' contribution beyond the simplest tasks it could devise. Alienated by the displacement of their interest from industry, workers fled, resisted, rebelled, and endured. Permitted only minimal contribution to the production process, and minimal benefit from it, they offered declining loyalty, effort, and interest to it. Such a style of operation makes it unlikely that industry will draw from workers their exertion, while suppressing their potential efforts of mind and skill; unlikely that it will induce overseers to maximize production at the expense of their former cohorts and daily associates; or unlikely that it will find any but the highest-level employees dedicated to the owners' interests. Future production envisions robots capable of performing only assigned tasks. Current production asks of men and women input as limited as that of a robot, yet those very workers possess the wealth of abilities and understanding that a robot can scarcely mimic.

The workers at the Boott never accepted the system that has been described; in fact, every group left. While "almost all Americans saw their standards of living increase," as Reich points out, these people did not. Nor was their treatment incidental or accidental; it was part of a pervasive strategy of operation. Owners at the Boott wanted to diminish labor's share of revenues to reduce costs, reserving the benefits of improvements in technology and organization to themselves, but with increasing expectations of exertion. The Boott and others rapidly became extractive industries, deriving profit from labor with declining recompense and shrinking investment.[8]

Many now agree that the status quo no longer works for American industry. As the MIT Commission on Industrial Productivity study *Made in America* details, methods that produced success in the good years—the pursuit of "interchangeability of workers with limited skill," low wages, and confrontational opposition to unionization—produced success through "rapacious" exploitation rather than development of a workforce representing an asset to be tapped rather than "a cost to be controlled." The tradition deriving from such conduct has led to lowered productivity among white- as well as blue-collar employees. Furthermore, "traditional manufacturing strategies, aimed at using capital to replace and to control labor, have the double disadvantage of requiring expensive capital and stripping out of the production process what must in the end be a distinctive American advantage—a skilled and educated labor force." This strategy also eliminated labor's ability to consume the products of manufacture, as Lowell so aptly demonstrated during its post-textile depression.[9] If one of the country's goals is, or should be, economic security for its citizens, it is important to note that textile workers never had it.

The current widely noted malaise affecting American industry, particularly as highlighted by competition in world markets, invites us to consider the conduct and lessons of the Boott Mill, our industrial past, in terms of the conduct or design of our industrial future. It is no coincidence that the entire industrial relations system is in turmoil while the country struggles to establish or re-establish its economic position on the world stage. Management, it is suggested, has failed to adjust to the new world of competition.[10]

From the time of Nathan Appleton to that of Frederick Taylor (and after) owners have noted their desire for their factories to be smoothly running machines. The Boott's reliance on the talent of its managers and workers for success stands out. Lack of regular expenditures on buildings or equipment throughout its history left the operators, both labor and oversight, as the crucial factor in its perpetuation. Yet despite the profitability of this style, despite of the power of the owner's organization and their control of the city, the capital generated in Lowell was invested elsewhere. Such a development augured badly for the future of an industry in any community. As workers at the Boott learned, concessions did not keep the plants open. They offered only temporary continuation of the pattern of rising loads and falling pay.

Because we find the same issues of definitions of tasks and amounts and speeds of work a primary issue in all industries, whether in relation to computer-controlled machine tools, automobile assembly lines, meat-packing houses, steel mills, and others, the issue is not one of textiles but of industrial work in the United States. Industry creates the situation that workers object to, flee from, and contribute less than their best efforts to. If people are asked to do less and less of a task as it is divided to reduce skill requirements, they are unlikely to bring the broad range of their capabilities to bear on the lesser assignments. If a worker can run a machine, make adjustments and repairs, innovate in terms of practice and technology, and improve product, then that person is frustrated when reduced to monitoring the process.

Faced with the end of the Cold War by which national aspirations had been defined and success measured, confronted by the relative lack of success in the changing world economy, the United States may be as adrift as the former Eastern Bloc countries. Providing solutions to the economy's problems has become a new industry unto itself. Some expect that the more educated members of society will ultimately profit from the global economy, and call for no regulation of it, while others see such regulation as a primary need. The information available from computer technology strikes some as the path to escape from conflict within production, while others see a new government-business partnership as the change which will ease our way. Increased personal savings, tax reform, and other changes have been proposed. One source lists eleven schools of thought in one paragraph.[11] A theme of "unity" runs through much of this literature, yet unity is impossible with a freebooting style of capitalism running rough-shod over labor, free from moral or cultural or governmental rein, with the world as its playing field. Current diagnoses of and prescriptions for the economy range from seeking "a new lead industry" for New England (textiles and computers having passed), to restoring to industry an (illusory) earlier long-range view that produced lasting success, to devising new strategies based on foreign models. Yet these do not address the problems arising from the path textiles blazed and others followed, from the early days of joint-stock corporations to recent years of runaway industry.[12]

Owners and workers do not, and correctly perceive that they do not, share common goals in our system of production. Some of the original Boott workers fled the mill as soon as they entered it. The entire labor force of native-born women left once the system was speeded up and stretched out beyond the loads they had once accepted. Workers continually rejected the directed technological and labor practices, and one after another the people filling the factory left it. Only the repeated waves of immigration perpetuated this style of operation.

Solutions must address the nature of the relationship between ownership and labor, and in turn that between industry and its environment. Industry must not only avoid creating problems in communities, but also reverse the policies which subjugate, diminish, and alienate labor. Resolving the contest between ownership and operation would obviate the need to minimalize tasks, to diminish labor (whether in terms of its cost per person or its control). If industry had not developed in a context of an ever-renewing supply of workers and the consistent opportunity to relocate to nonindustrial areas, these issues would have to have been confronted long ago.

Management at the Boott did not behave improperly, according to the code of business behavior, nor was it an unsuccessful enterprise for its owners. The principals were not necessarily unkind or ungenerous. The Flathers played leading roles in the affairs of such worthy institutions as the YMCA, Boy Scouts, and their church. Director Frederick Fanning Ayer sent large sums to local charities from his New York home.[13] Yet when the corporation was closed down in 1954, Flather, a minor stockholder, received $229,470 for stock he had acquired for $123,307, representing a profit of $106,163 in addition to the dividends and family salaries received previously. Workers received dismissal slips. (In November 1954, someone in the office marked twelve of fourteen names on a union ballot "Terminated" or "Laid-off," just two "Working."[14]

In a country divided by race, class, and numerous political issues, we are largely united in our relationship to the changing world order: we will change our economy, or we will suffer. How we manufacture will matter to all of us. It mattered to the Boott employees, both workers and managers, and therefore they kept the place going. But two problems interfered, fatally: (1) the manufacturing process didn't matter much to the owners, who continued to benefit from income generated by this and other operations, and (2) the employees saw that and recognized that their efforts to manufacture were frustrated, not fulfilled. That ownership saw the dichotomy of interests was demonstrated by their continuing search for ways in which to pay bonuses to even the highest-level workers to induce increased production. The workers' recognition of it appeared in their comments and actions: they complained and left. High-tech, high-skill, flexible production, niche markets—none of these will alter the basic situation.

Many writers find current conditions puzzling and worrisome. Others see opportunity: "We see the current moment as one of those historical periods of transformation in which existing institutional structures have been challenged and opened up to experiment in ways that allow considerable choice in how to reconstruct and modify them to best serve the interest of worker, employers, and society in general."[15] Such a task seems daunting, and the appetite for it among politicians, invisible. Yet as a subject for national attention, the nature of the relationship between owners and workers "is equally if not more important than the family, the school, the farm, the stock market, or the other settings in our domestic life that have figured much more prominently in recent political debates. What goes on in the employment relationship makes a huge difference in the economic and social fate of the vast majority of American citizens, as well as in the productivity and competitiveness of American enterprise."[16] The issue isn't simply who governs the workplace, but who governs the course of industry. That question is too important to be left to stockholders, who have the least interest in an overall plan, as long as there is income from some source, some location. The great numbers of people, the employees and other citizens, have the most interest. The relationship between owners and employees in this country has long been oppositional. The time and energy robbed from productivity by this conflict could go far to fuel recovery, if that energy became available. Excessive numbers of unproductive managers, feather-bedding, mutual deception, and de-skilling all prevent rather than foster productivity. The goal of industry, "diminished jobs," results in "diminished people."[17]

The history of owners and workers appealing to public bodies to further their separate interests is as old as industry. Owners found both benefits and costs associated with government's role. Business benefited from social constructs favoring it: "Patents, trademarks, the institution of bankruptcy, the corporation itself, even the concept of legally enforceable private property, are all political inventions."[18] The need now is for (1) a course that will advance the interest of the mass of people by changing their place in and relationship to the companies for which they work; and (2) a national strategy of competitiveness in the global economy which can take advantage of the new situation. As Kuttner puts it, "industrial America needs a new form of social partnership. . . , a new social contract between labor and management."[19] Ideally, this would represent a partnership of all contributors to production: labor, management, ownership, and society.

Such a policy would have significant impacts on how industry operates within the country. It could remove the spectacle of states bidding against one another on the basis of diminished expectations to attract firms. Why should people in one state have to offer lower wages, poorly funded education, less safety, less security to compete with those in another? A new social contract would create mutual obligations by giving workers a voice and a stake in the improvement of their companies. Why would employees strive to increase productivity when to do so does not lead to raises and may even cost them their jobs? A new policy could also aim to produce mutual goals and procedures to dispense with the wasteful confrontation between the parties to production. If we ever could, we can no longer afford a system in which everyone looks out for number one; we need agreement on how to make Henry Miles' identity of "public morals [or goals] and private interests" a reality. Given such changes, workers could become maximum contributors to production and major beneficiaries from it, rather than an enemy to be subdued, controlled, and eliminated. Sufficient involvement by all parties would enable them to contribute to mutual interests in income, security, quality, and productivity. Such developments would require "increased trust, commitment, and cooperation at the workplace rather than further institutionalization of adversarial relationships" from all concerned.[20]

In recent years the United States has slipped back from previous levels of economic growth as measured by the balance of trade and expectations of improvement in the standard of living for most people. Even though productivity has risen steadily over the past decade, that share which used to go to labor no longer does. Not simply international competition, but the political climate that weakens labor and strengthens business's ability to define the conditions of employment have produced this situation.[21] Without fundamental change, few observers expect this trend to stop. While many factors can contribute to the necessary changes, it appears to me that the division between owners and workers is the most fundamental, the most wasteful in its draining away of potential contributions and accomplishments.

If a person lacks the opportunity for meaningful, productive work, then that person's talents are wasted, not put to society's use, and we are all the poorer, our economy and culture the weaker, for it. In order to become as strong as we can be, as strong as we need to be to compete successfully in the new world order, all people have to have the opportunity to develop their talents and express that which they can offer. They must have a role in society and a stake in their employment. Currently workers called managers have some, but not all, of these characteristics: they receive and utilize education, make decisions, plan, accomplish goals, and may benefit from their performance. Other workers generally fail to find these attributes in their jobs and, correctly not anticipating these opportunities (let alone accompanying rewards), have little incentive to pursue the education which most writers identify as one of the needs for American success. We can't change one without the other. More hours in school will not educate those without opportunity to exercise their learning, as it is plain that many students already recognize, seeing no reason to study that which they will have little or no opportunity to apply.

The Boott Cotton Mills worked for its owners from start to finish, but it worked poorly for all but the top level of employees. In the midst of intensifying international economic competition, such a style of operation does not serve the interests of the country or its working people, either labor or (most of) management. Managers, workers, and most other citizens must recognize their common interests and identify and pursue their common goals.

Note on Sources

In this work, I have drawn upon a variety of sources traditional to such histories, many of which are cited in the notes, but I have also benefited greatly from some not available or little used previously. Most notably, the Flather Collection, given to the Lowell Museum and available at Lowell University, offers voluminous material on the career of Frederick A. Flather and the operation of the Boott Mills, 1905–55. It also contains very important records created by John Rogers Flather during his years at the Boott. Without this collection, my book would not have been possible. Without the cooperation of Frederick Flather, since deceased, and access to his personal collections, this book would have been much less complete, its assessments much less certain. Oral histories conducted by the Lowell National Historical Park and others associated with the university also made significant contributions. I am grateful to the donors, the sharers, the compilers, and the contributors.

Notes

Introduction

1. Rev. Henry A. Miles, *Lowell, As It Was, And As It Is* (Lowell, 1845), p. 128.
2. Thomas Dublin, *Women at Work: The Transformation of Work and Community in Lowell, Massachusetts, 1826–1860* (New York, 1979), p. 9.
3. Alfred D. Chandler, Jr., *The Visible Hand: The Managerial Revolution in American Business* (Cambridge, 1977), pp. 4–5, 6.
4. David Brody, *Workers in Industrial America: Essays on the Twentieth Century Struggle* (New York, 1980), p. 6.
5. Caroline F. Ware, *The Early New England Cotton Manufacture: A Study in Industrial Beginnings* (Boston, 1931), p. 103.
6. James H. Green, *World of the Worker: Labor in Twentieth-Century America* (New York, 1980), p. 84.
7. Vera Shlakeman, *Economic History of a Factory Town: A Study of Chicopee, Massachusetts* (New York, 1969 [reprint of 1934 edition]), p. 138.
8. Green, *World of the Worker,* p. 84; Philip Scranton, *Proprietary Capitalism: The Textile Manufacture at Philadelphia, 1800–1885* (Cambridge, 1983), pp. 355–56; Brody, *Workers,* p. 206.
9. Constance McLaughlin Greene, *Holyoke, Massachusetts: A Case History of the Industrial Revolution in America* (New Haven, 1939), p. 64.
10. Scranton, *Proprietary Capitalism,* p. 33.

Chapter One. Lowell to 1850

1. George Sweet Gibb, *The Saco-Lowell Shops: Textile Machinery in New England, 1813–1949* (Cambridge, 1950), p. 23. Numerous scholars have addressed the development of the United States textile industry; perhaps the best single bibliography is the one compiled by Martha Mayo in Louis Eno, ed., *Cotton Was King: A History of Lowell Massachusetts* (New Hampshire, 1976).
2. Gibb, *Saco-Lowell,* pp. 65–66.
3. Robert F. Dalzell, Jr., *Enterprising Elite: The Boston Associates and the World They Made* (Cambridge, 1987), p. 67.
4. Boott Cotton Mill (BCM) Indenture, copy, private collection (PC).
5. See my "Building on Success: Lowell Mill Construction and Its Results," *Journal for the Society for Industrial Archeology* 14, no. 2 (1988): 23–24.
6. Laurence F. Gross and Russell A. Wright, "Historic Structure Report, History Portion: Building 6, the Counting House, the Adjacent Courtyard, and the Facades of Buildings 1 and 2; Boott Mill Complex, Lowell National Historical Park, Lowell, Massachusetts" (HSR), prepared for National Park Service, Denver Service Center, Denver, Colo., pp. 56, 79–82; and the Stenographic Record of Boott Mills, Appellant, v. Board of Assessors of the City of Lowell, Appellee, Before Appellate Tax Board, Boston, February 26, 1945, 1:60 (hereafter, Stenographic Record), PC.

7. Gibb, *Saco-Lowell,* Appendix 1.
8. HSR, p. 56; Agent's Letterbook, p. 209, University of Lowell, Special Collections.
9. Arthur T. Safford to Frederick A. Flather (FAF), August 13, 1947, p. 2, PC. For the most thorough available description of Lowell's waterpower development, see Patrick M. Malone, *The Lowell Canal System* (Lowell, 1976).
10. Safford to FAF, pp. 2–5.
11. "Sketch of Boott Cotton Mills," draft prepared by BCM for a new edition of Lamb's *Cotton Industry of the U. S.* (James H. Lamb Co., Boston), p. 1, PC.
12. See, e.g., Nathan Appleton, *Introduction of the Power Loom and Origin of Lowell* (Lowell, 1858).
13. Robert Means to Amos Lawrence, 5/18/36, Amos Lawrence Papers, Letterbook 2 (395), Massachusetts Historical Society; I am indebted to Duncan Hay for this reference.
14. Andrew Ure, *The Philosophy of Manufactures: or, An Exposition of the Scientific, Moral, and Commercial Economy of the Factory System of Great Britain* (London, 1835), pp. 15–16.
15. Frances W. Gregory, *Nathan Appleton, Merchant and Entrepreneur, 1779–1861* (Charlottesville, 1975), p. 191; Nathan Appleton, quoted in Thomas Bender, *Toward an Urban Vision: Ideas and Institutions in Nineteenth-Century America* (Lexington, c. 1975), p. 99.
16. Miles, *Lowell,* p. 128.
17. *Report of the Special Commission on the Hours of Labor and the Condition and Prospects of the Industrial Classes* (Boston, 1866), p. 13.
18. Personal communication.
19. B. F. French, Agt., to William Boott, Esq., 3/22/1844, Massachusetts House of Representatives, Document #50, 1845, p. 24; Miles, *Lowell,* pp. 75–76.
20. Both Paul Hudon and Scranton have been helpful in developing this point. The latter suggests that the common phrase for employment, to be "on the corporation," may reflect Lowellians' recognition of the new situation. Both Anthony Wallace, in *St. Clare: A Nineteenth-Century Coal Town's Experience with a Disaster-prone Industry* (New York, 1987), p. 273, and Paul E. Johnson, in *A Shopkeeper's Millennium: Society and Revivals in Rochester, New York, 1815–1837* (New York, 1978), discuss the owners' divestiture of responsibility for labor.
21. Quoted in Shlakeman, *Chicopee,* p. 147.
22. French to Boott, 3/22/1844, p. 22.
23. J. Metcalf to Joseph C. Metcalf (son), 8/30/1819, Museum of American Textile History (MATH).
24. Sarah and Charles to Mrs. Metcalf, 4/5/1843, MATH; Dublin, *Women at Work,* p. 42; Ware, *Early New England Cotton Manufacture,* p. 255.
25. Sarah and Charles Metcalf, 4/5/1843, Charles 4/27/1844 (seem to be same year, probably 1844), MATH.
26. Mary to Mrs. Metcalf, 11/14/1847, 12/12/1847, 1/23/1848, MATH.
27. Johnson, *A Shopkeeper's Millennium,* describes the religious revivals of the early nineteenth century as serving "not the needs of 'society' but of entrepreneurs who employed wage labor," p. 137. Similarly, he describes drinking as part of the "pattern of irregular work and easy sociability sustained by the household economy," preceding the factory system, p. 57.
28. Regulations signed by one Mary P. Johnson, presumably another employee

of the day, but one of the many evidently leaving no record, PC; for lack of enforcement see, e.g., Dublin, *Women at Work*, p. 60.

29. Mary to Mrs. Metcalf, 12/12/1847, 5/21/1848, MATH; when water in the river was high, it flooded the wheelpits from below the mills and retarded the wheels, which had to turn through it.
30. Mary to Mrs. Metcalf, 2/12/1848, MATH.
31. Gibb, *Saco-Lowell*, pp. 70, 89.
32. For use of sailors' wages as mercantilists' economic shock absorber, see Marcus Rediker, *Between the Devil and the Deep Blue Sea: Merchant Seamen, Pirates, and the Anglo-American Maritime World, 1700–1750* (Cambridge, 1987), p. 145; "Report of Committee Appointed at Annual Meeting, Boott Mills, Mar. 6/43 to Investigate the Affaires of the Company," unpaginated, manuscript 442, Baker Library.
33. Dalzell, *Enterprising Elite*, p. 67.
34. George F. Kenngott, *The Record of a City: A Social Survey of Lowell, Massachusetts* (New York, 1912), pp. 12–15.
35. Gregory, *Nathan Appleton*, p. 196.
36. Merton M. Sealts, Jr., ed., *The Journals and Miscellaneous Notebooks of Ralph Waldo Emerson* (Cambridge, 1973), 10:102.
37. Ibid., pp. 102–3.
38. In Hannah Josephson, *The Golden Threads: New England's Mill Girls and Magnates* (New York, 1949), p. 252.
39. See Metcalf Letters, MATH; Preston Letters, private collection; Jonathan Prude, *The Coming of Industrial Order: Town and Factory Life in Rural Massachusetts, 1810–1860* (Cambridge, 1983).
40. In Josephson, *Golden Threads*, p. 116.
41. Helena Wright, "Sarah G. Bagley: A Biographical Note," *Labor History* 20 (1979): 409 and note.

CHAPTER TWO. A LOSS OF LUSTRE, 1850–1870

1. Ware, *Early New England Cotton Manufacture*, p. 188.
2. Dublin, *Women at Work*, pp. 138–39, 140–41, 198.
3. Ibid., pp. 165, 192.
4. Ware, *Early New England Cotton Manufacture*, p. 113.
5. 1864 Time Table, PC; winter hours for December and January simply shifted the schedule to 7:00 to 7:00; B. F. French to William Boott, 1844, p. 22.
6. "Wage and Cost Rates, 1864–1893," box 29, Flather Collection (FC), Lowell Museum (at Lowell University), p. 43.
7. Steven David Lubar, "Corporate and Urban Contexts of Textile Technology in Nineteenth Century Lowell, Massachusetts: A Study of the Social Nature of Technological Knowledge," Ph.D. diss. (University of Chicago, 1983), p. 121; Lubar allots them just $2 per day at this time; Henry B. Call, Tr., to John Wright, Agt., Suffolk, 4/22/1850, box 29, FC; Dublin, *Women at Work*, p. 192; *Statistics of Labor:* "Wages and Prices, 'Cotton Goods'" (Boston, January 1879), pp. 70–71.
8. Robert G. Layer, *Earnings of Cotton Mill Operatives, 1825–1914* (Cambridge, 1955), pp. 46–47.
9. Ibid., p. 45.

10. Philip Scranton, *Figured Tapestry: Production, Markets, and Power in Philadelphia Textiles, 1855–1941* (Cambridge, 1989), p. 8.
11. Dublin, *Women at Work,* pp. 71–72, 103, 110–11, 203.
12. Melvin T. Copeland, *The Cotton Manufacturing Industry of the United States* (Cambridge, 1912), p. 157.
13. Henry B. Call to John Wright, 1850, Lowell Historical Society; Dalzell's contention (*Enterprising Elite,* pp. 54–55) that output and wages were not controlled in Lowell is contradicted by this and other such evidence.
14. James B. Francis to T. Jefferson Coolidge, 12/9/1860, PC; Dalzell shows how they used Massachusetts Hospital Life Insurance Company in the same way.
15. F. B. Crowninshield to Coolidge, 1/21/1862, PC.
16. Series of notices from the Arbitrators, including Francis, to Coolidge, PC; E. E. Manton, Secretary (by Wm. B. Whiting), Office of the Boston Manufacturers' Mutual Fire Insurance Company, July, 1864, account of 1864 Salmon Falls Mill fire, PC; a danger inherent in such close circumstances appeared in a letter from Arthur T. Lyman to Coolidge (n.d., PC): "I am glad to learn that the reports about Mr. Child which I heard from Mr. Francis and others are unfounded and I will have them contradicted." Neither the ties of association nor the dignity of the personages could prevent gossip, it appears (sources do not reveal its nature).
17. Barry Bluestone and Bennett Harrison, *The Deindustrialization of America: Plant Closings, Community Abandonment and the Dismantling of Basic Industry* (New York, 1982), p. 134.
18. J.[?] Bartlett to Boott Mill, 5/7/63; S. G. Newton to Coolidge 7/17/63; box 29, FC.
19. James C. Ayer to Coolidge, 12/26/60, box 29, FC.
20. J. C. Ayer, M.D., *Some of the Uses and Abuses in the Management of Our Manufacturing Corporations* (Lowell, 1863, pp. 18–24); similar complaints regarding the selling house had appeared in the Boott's *Internal Report* (1843).
21. Ayer, *Uses and Abuses,* p. 9.
22. 1859 letter, PC.
23. Lubar, "Textile Technology," pp. 79, 94, 82, 122, 133, 231.
24. J. T. Stevenson to Coolidge, 1/21/1859 (?), box 29, FC; Copeland, *Cotton Manufacturing,* later (p. 63) claimed that, "By 1960, speeders had been largely superseded in American mills by fly-frames"; Gibb, *Saco-Lowell,* refutes this claim, p. 193.
25. *Report to the Directors of the Boott Cotton Mills, October 22d, 1860,* and *April 26, 1861,* PC; ensuing material is based on these reports.
26. Chandler, *Visible Hand,* pp. 247 and 71.
27. Ibid., p. 68.
28. [Illegible] to Coolidge, 7/10/1861, PC; Lubar, "Textile Technology," cites similar attitudes, pp. 52–53.
29. In contrast, an attack by Ayer on the operation of the Middlesex Mills noted that "the property is reported to be in excellent order and condition, the buildings in good repair, and the machinery equal to new in point of service." Lack of maintenance had apparently taken its toll in more lasting ways at the Boott than at the Middlesex.
30. S. J. Whethrill (?) to Coolidge, 4/6/1860, PC.
31. The utility of staff advice could, of course, be limited by either lack of expertise or lack of confidence in loyalty to the Boott, rather than to a supplier.

32. T. J. Coolidge, "Report to the Directors of the Boott Cotton Mill, 10/22/1860," PC.
33. Copeland, *Cotton Manufacturing*, pp. 262–63.
34. James B. Francis to Linus Child, Esq., 10/8/1860, PC. The Warren was a simple scroll-case turbine made by the American Water Wheel Co., of Boston, of which A. Warren was agent and presumably the wheel's designer; Warren to Linus Child, 3/13/1860, PC; see also *Transactions of the American Society of Civil Engineers* 85 (1922): 1256.
35. John Aiken to Coolidge, 2/15/1864, PC.
36. Various letters, PC; *Boott Cotton Mills Trial Balance*, 11/30/1865, PC.
37. Charles Cowley, *Illustrated History of Lowell* (Boston, 1868), p. 60.
38. Geo. Stark, Mngr. Boston & Lowell and Nashua & Lowell railroads, to Coolidge, 10/30/1861, PC.
39. Exchange of letters between Coolidge and Geo. H. Kuhn, J. Huntington Wolcott, Geo. A. Gardner, F. B. Crowninshield, and J. A. Lowell, 5/22/1862–6/26/1862, PC.
40. Lubar, "Textile Technology," p. 73.
41. Ayer, *Uses and Abuses*, p. 10.
42. For Philadelphia, see Scranton, *Proprietary Capitalism*.
43. James G. ? to Coolidge, 12/9/1861; George Dexter to Coolidge, 11/17/1865, PC.
44. Letters to Coolidge from Edmund Dwight and George Dexter, 12/10/1862, 12/23/1862, 12/9/1865, PC.
45. Dexter to Coolidge, 2/20/1865, PC.
46. Memorandum of Agreement, ca. 11/2/1869, signed by eight treasurers representing eleven companies, PC.
47. E.g., Gibb, *Saco-Lowell*, p. 198. On other occasions comparable construction work took place during operation.
48. See Dalzell, *Enterprising Elite*, pp. 222–23.

CHAPTER THREE. ORGANIZATION AND EQUIPMENT

1. Boott Cotton Mills (BCM) Correspondence with American Trading Company, 1888, FC.
2. For details of expansion, see Gross and Wright, HSR, pp. 162–95.
3. 1882 *Barlow Insurance Survey* of the BCM, Museum of American Textile History (MATH).
4. Samuel Webber, *Manual of Power for Machine, Shafts, and Belts, with the History of Cotton Manufacture in the United States* (New York, rev. ed., 1879).
5. "Report: Lowell National Historical Park and Preservation District: Cultural Resources Inventory," prepared for the Division of Cultural Resources, North Atlantic Office, National Park Service, by Shepley, Bulfinch, Richardson, and Abbott, "Boott Cotton Mills" (1980), p. 18.
6. The wheels generated more power than previously because a greater volume of water was available to turn them.
7. Arthur T. Safford to Frederick A. Flather (FAF), p. 5, 8/13/1947.
8. Agent's Letterbook (Cumnock's), 10/26/1888–7/14/1891, p. 799; site maps; *Transactions of the American Society of Civil Engineers* 85 (1922): 1256.
9. Dalzell, *Enterprising Elite*, p. 226.

10. Papers related to piecework wages, 1889; V. I. Cumnock, "Memo," 1/14/1895, box 29, FC.
11. See *Davison's Textile Directories* for the 1890s, as well as bills and receipts, box 29, FC, and Letterbook, to Lincoln Park Co., Patterson [sic.], N.J., p. 362.
12. Walter E. Parker, "Report on the Condition of the BCM, Lowell, Mass., 1902" (n.p., 10/1902), FC, box 29, p. 26.
13. Ibid., pp. 25–26.
14. Scranton, *Figured Tapestry,* p. 386.
15. Parker, "Report," pp. 33–35.
16. Alexander Keyssar finds the shift from production for inventory to production to order a general event in the late nineteenth century: *Out of Work: The First Century of Unemployment in Massachusetts* (Cambridge, 1986), p. 16.
17. Anon., *Alexander Goodlet Cumnock, 1834–1919* (Lowell, 1953), p. 4.
18. Lubar, "Textile Technology," p. 82.
19. Letterbook, 8/16/1882; 4/23/1890.
20. Elliott C. Clark to A. G. Cumnock, 4/12/1888, box 29, FC.
21. Letterbook, p. 410, 1/6/1890.
22. Letterbook, p. 864.
23. Letterbook, pp. 920, 161.
24. Letterbook, p. 875, 3/19/1891, p. 937, 6/4/1891.
25. Letterbook, p. 908, 4/20/1891.
26. Letterbook, p. 474, J. G. Marshall to Clark.
27. Letterbook, p. 338, Marshall for Cumnock to "Bro. Israel," 10/20/1889.
28. Letterbook, pp. 527, 956.
29. Lubar, "Textile Technology," pp. 80–81; Chandler, *Visible Hand,* p. 68; see Chapter 1, above.
30. R. G. Dun, Manuscript Field Reports, vol. 51, pp. 668, 673, 897, 891, Baker Library.
31. *Cumnock,* p. 5.
32. Lowell Machine Shop Records (LMS), Baker Library, HB-2, pp. 1–26.
33. [Capt. S. K. Hutchinson], "Valuation of Lowell Mills for Mutual Insurance" (1873 manuscript, n.p.), FC.
34. Gibb, *Saco-Lowell,* p. 757 n.
35. Ibid.
36. Cumnock, "Proceedings," *New England Cotton Manufacturers Association* (NECMA) 15 (10/15/1873): 31–34.
37. In 1871, Jacob H. Sawyer, agent of the Appleton Mills in Lowell, addressed the New England Cotton Manufacturers Association and described the new spindle he had tested at the Boott. Through observation and experimentation he had developed a spindle that was supported at its center of gravity, ran steadily, and could be run at up to 7,500 rpm. "So unwavering was the operation of the faster and lighter spindle that even at its higher speeds it produced less breakage and more uniform yarn than had been possible with the common spindle." Samuel Webber tested the frames using the new spindle in the Boott in August 1876 and recorded their performance: 5 1/2–oz. spindles ran at 6,020 rpm, front rolls turned at 97 rpm producing a draft of 8.80 and requiring 2.144 horsepower per frame. These figures not only recorded the operation of the frames but also indicated, again, the precision and sophistication with which machinery could be tested and evaluated.

38. Thomas R. Navin, *The Whitin Machine Works since 1831: A Textile Machinery Company in an Industrial Village* (Cambridge, 1950), p. 109.
39. Channing Whittaker, *Merrimack, Tremont, Suffolk, Boott,* notebook, Lowell University Special Collections.
40. A. G. Cumnock, "Proceedings," *NECMA* 27 (10/29/1879): 60; Webber, *Manual of Power,* pp. 88–90. The "science" of shafting had made significant advances as implemented in Lowell over the years. Paul Moody had helped devise a system that substituted belts for gearing in connection with shaft drive during the earlier years, and subsequently engineers had furthered the system's sophistication. James B. Francis had continued the experimentation and in 1867 published the results of his studies in the *Journal of the Franklin Institute.* The complexity and specificity of his calculations provided a mathematical basis for determining the necessary diameter of various shafts according to the amount of power to be transmitted and speed of revolution. His precise formulas remind us that the engineering of these mills drew upon sophisticated theoretical foundations.
41. Gibb, *Saco-Lowell,* p. 193.
42. Webber, *Manual of Power,* p. 54, part 2. Up until the Civil War, United States manufacturers relied largely on the Sharp and Roberts pattern mule, of the style of the English inventor of these machines, Richard Roberts, and produced by the Mason Machine Works of Taunton, Massachusetts. The Civil War's onset took domestic machine-builders out of the textile-machine business to a great extent, and manufacturers began to import two English mules, one by Parr, Curtis, and Madely and another, the one adopted by the LMS after the war, by Platt Brothers of Oldham, Lancashire, "both, in all essential features, lineal descendants of the Sharp and Roberts mule."
43. Lowell Machine Shop Records, OH-1, Baker Library.
44. Each bar in the chain corresponded to one pick; each bar carried, for each harness in use, a riser or a sinker. The former lifted levers and caused the corresponding harness to be raised, while the latter caused the remaining harnesses to be lowered.
45. If one color predominated in the pattern, then several shuttles would carry that color and serve a filling-mixing purpose. (While all yarn may look pretty much the same to the untrained eye, even identical counts of yarn spun on different machines or from different batches of cotton varied slightly; by intermixing several bobbins of yarn in the filling, these differences were minimized [compared to the effect that would be produced if one bobbin traveled back and forth in one section of cloth until it was empty].)

    The drop-boxes were controlled by the same type of chain as the harnesses. Before each pick the chain indicated to the boxes, through a series of cams, gears, levers, and chains, which shuttle was to be thrown. The operation of these box motions, of course, had to be very sure and precise. Any misalignment of the box and the race plate would result in the shuttle being thrown out of the loom at great speed and considerable danger.
46. In addition to these features that made the loom more "automatic," certain other advances played a part in the novelty of the machines. The yarn on each bobbin was attached to the battery in order that it would begin to trail across the loom when inserted. The temple on the right side of the loom was equipped with a pair of blades which cut off the resulting trailing ends. The take-up roll was also altered. Instead of the cloth passing over the

breast beam and then down onto the take-up, a high take-up roll behind the breast beam guided the material to a cloth roller directly beneath. This set-up left more room for a fixer to work and also kept the cloth at a more consistent width by locating the take-up closer to the fell of the cloth. The Draper automatics were also novel in that they were always made one-handed, with the shipper handle on the left and the battery on the right. Previously looms had been ordered in right- and left-handed pairs to concentrate drive pulleys near bearings on the line shafts of the power system.

CHAPTER FOUR. THE INTERACTION OF PEOPLE AND MACHINES

1. William A. Burke, "Proceedings," *NECMA* 21 (10/25/1876): 15–16.
2. Statistics are from the annual *Statistics of Lowell Manufactures.*
3. Donald B. Cole, *Immigrant City, Lawrence, Massachusetts; 1845–1921* (Chapel Hill, 1963), p. 235.
4. T. M. Young, *The American Cotton Industry: A Study of Work and Workers, Contributed to the Manchester Guardian* (New York, 1903), p. 4.
5. Cumnock, "Proceedings," *NECMA* 5 (10/20/1868): 23–24.
6. Cumnock's attention to the oil's "flash point" related to the issue of fire protection. Cumnock, "Proceedings," *NECMA* 26 (4/30/1879): 15–16.
7. From *The American Wool and Cotton Reporter,* quoted in *Cotton and Rayon Textile Companies: An Analysis of Comparative Investment Values and Opportunities,* clipping, FC.
8. "Fire Regulations," Proprietors of Locks and Canals Collection, inside front cover, DC-2, Baker Library.
9. Parker, "Report," p. 10.
10. Ibid.; *American Wool and Cotton Reporter* 17 (1903): 522.
11. Letterbook, p. 453, Cumnock to Rufus R. Wade, Chief, Mass. District Police.
12. Kenngott, *Record of a City,* pp. 141–42, p. 146.
13. Letterbook, p. 277, 8/22/1889, Cumnock to J. C. Howe.
14. Accident statistics and quotes, unless otherwise noted, are from copies of reports in the Agent's Letterbook for 10/26/1888–7/14/1891, Lowell University.
15. Letterbook, pp. 312, 315, 9/17/1889.
16. Letterbook, p. 569, 5/27/1890.
17. Letter from American Mutual Liability Insurance Co. of Boston, Mass., to BCM, 8/18/1888.
18. Letterbook, p. 628, 8/1/1890; p. 572, 6/2/1890.
19. Letterbook, p. 668, 8/23/1890; p. 624, 7/30/1890; p. 81, 1/10/1889.
20. Letterbook, p. 254, 8/21/1889; p. 423, 1/18/1890.
21. Letterbook, p. 317.
22. Letterbook, p. 764; p. 876, 3/19/1891.
23. Letterbook, pp. 485–86.
24. Letterbook, p. 976; p. 44, 12/18/1888.
25. Letterbook, p. 838, 2/18/1891, p. 851, 3/2/1891; p. 556, 3/19/1890; p. 728, 11/3/1890; pp. 301, 303.
26. Letterbook, p. 748, 11/17/1890; p. 867.

27. Letterbook, p. 463, 2/26/1890; pp. 673, 676, 8/23/1890, p. 686, 9/26/1890; pp. 896, 869, 3/17/1891; p. 139, 3/21/1889; p. 141, Letter from American Mutual Liability Insurance Co. of Boston, Mass., to BCM, 3/27/1888, FC.
28. Parker, "Report," p. 10.
29. F. A. Leigh, "Repairs in Cotton Mills," *NECMA* 19 (10/27/1875): 31.
30. Payroll records for this period are far from complete, but wages for most of the jobs associated with the operation have been uncovered.
31. "Five Weeks Ending Saturday Evening, November 7, 1874," box 29, FC.
32. Young, *American Cotton Industry*, p. 32.
33. Herbert J. Lahne, *The Cotton Mill Worker* (New York, 1944), cited by Dublin, *Women at Work*, p. 143; wages and prices, box 29, FC.
34. Keyssar, *Out of Work*, p. 404 n; *Statistics of Labor*, 1879, pp. 78–79.
35. Layer, *Earnings*, p. 45.
36. Fidelia O. Brown, "Decline and Fall: The End of the Dream," in *Cotton Was King*, Arthur L. Eno, ed. (Lowell, 1976), p. 148.
37. Layer, *Earnings*, pp. 38ff.
38. Parker, "Report," p. 10.

CHAPTER FIVE. LABOR AND MANAGEMENT

1. Copeland, *Cotton Manufacturing*, pp. 112–13.
2. Letterbook, p. 810, Cumnock to Clark, for Mr. Wadlen, Massachusetts Bureau of Labor, yr. end 11/1/1890.
3. Copeland, *Cotton Manufacturing*, p. 115.
4. Dublin, *Women at Work*, p. 143.
5. Figures from *Statistics of Labor* (Feb. 1878), Public Document no. 31, pp. 237–38.
6. Letterbook, p. 107, Cumnock to Clark, 2/6/1889.
7. Cumnock, "Proceedings," *NECMA* 18 (4/21/1875): 21–22.
8. Above and ensuing examples from the Agent's Letterbook unless otherwise noted: pp. 488, 518, 290, 913, also p. 283.
9. Letterbook, pp. 316, 88, 682; Green, *World of the Worker*, p. 101–102.
10. I am indebted to Philip Scranton for the "personalism" aspect of this interpretation; see also Jacquelyn Hall et al., *Like a Family: The Making of a Southern Cotton Mill World* (Chapel Hill, 1987), p. 92, for another example of this idea of bosses seeking loyalty.
11. Letterbook, p. 356, also pp. 387, 393, 401, 666.
12. Letterbook, p. 880; p. 942, 6/8/1891.
13. Letterbook, pp. 159, 947.
14. Letterbook, pp. 248, 131, 205, 258, 290, 525, 136, 963, 964, 714, 901.
15. Letterbook, pp. 534, 314, 794, 709–10.
16. Letterbook, pp. 492, 914, 910, 710, 902.
17. Copeland, *Cotton Manufacturing*, pp. 157–59.
18. Letters from Associated Textile Industries to BCM, 4/12/1888, and 3/24/1887, box 29, FC.
19. Cumnock to E. M. Shaw, Letterbook, p. 468, re *Lowell Citizen*, 2/27/1890.
20. Letterbook, pp. 360, 960, 215, 339; Hall, *Like a Family*, p. 95.
21. "Thirty Years Ago," clipping, PC; similar gift-giving was also found by Judith

A. McGaw, *Most Wonderful Machine: Mechanization and Social Change in Berkshire Paper Making, 1801–1885* (Princeton, 1987), p. 326, as well as by Scranton, in his study of Philadelphia (personal communication).

22. Shlakeman, *Chicopee*, pp. 187–88.
23. Carol Polizotti Webb, "The Lowell Mulespinners Strike of 1875," *Surviving Hard Times*, Mary H. Blewett, ed. (Lowell, 1982), pp. 11–20.
24. Parker, "Report," p. 40.
25. *Wades, Fiber and Fabric* (a trade newspaper) 3, no. 77 (1886): 195.
26. Copeland, *Cotton Manufacturing*, p. 123.
27. Shirley Zebroski, "The 1903 Strike in the Lowell Cotton Mills," in Blewett, ed., *Surviving Hard Times*, p. 45.
28. *American Wool and Cotton Reporter* 17 (1903): 522.
29. Ibid., pp. 308, 365, 419, 447, 475, 522, 675, 785, 792.
30. "New Wage Schedule," *Lowell Evening Citizen*, 1/10/1905, clipping, PC.
31. "Boott Directors Given Authority to Sell Mills," *Lowell Evening Citizen*, 1/10/1905, clipping, PC.
32. Memo, 1950s, PC.
33. "Boott Directors Given Authority," *Lowell Evening Citizen.*
34. Lockwood-Greene and Co., "Sketch Plans Showing Proposed Reorganization of Boott Cotton Mills, Lowell, Mass.," April, 1903, box 29, FC.
35. See, e.g., Ayer, *Uses and Abuses;* Robert W. Dunn and Jack Hardy, *Labor and Textiles: A Study of Cotton and Wool Manufacturing* (New York, 1931), p. 15.
36. Memo, 1950s, PC.
37. Scranton, *Figured Tapestry*, p. 86.
38. Memo from V. I. Cumnock, Supt., 1/14/1895, box 29, FC; Cumnock to Clark, Letterbook, p. 130.
39. Cumnock, "Proceedings," *NECMA* 18 (4/21/1875): 21–22; Kenngott, *Record of a City*, p. 202; see Dalzell, *Enterprising Elite*, and Margaret Terrell Parker, *Lowell: A Study of Industrial Development* (New York, 1940), on limited industrial development; David Montgomery, *The Fall of the House of Labor: The Workplace, the State, and American Labor Activism, 1865–1925* (Cambridge, 1987), p. 169.
40. Copeland, *Cotton Manufacturing*, p. 156; Young, *American Cotton Industry*, p. 3; see also Gibb, *Saco-Lowell*, and Navin, *Whitin Machine Works.*
41. Personal communication from Frederick Flather (FF).
42. Copeland, *Cotton Manufacturing*, pp. 264, 355; Lubar, "Textile Technology," p. 171.

CHAPTER SIX. DIRECTION OF THE NEW BOOTT MILLS

1. Boott Mill file, "1904–1907 Transfer," quotes Flather, PC.
2. "Transition File," p. 2, PC.
3. Clipping, *Lowell Daily Mail*, 2/16/1905, box 14, FC; FAF, Directors' Meeting account, Directors' Minutes (hereafter DM) 1/25/1941, PC; Frank E. Dunbar to Wellington, Sears 2/14/1905, PC; untitled document describing purchase agreement, FC.
4. Clipping, *Lowell Daily Mail*, 2/16/1905.
5. Letter from Lockwood-Greene to Boott Mill (BM), 1/29/1907, box 51, FC.
6. F. P. Sheldon to BM, 1907, pp. 3–4, box 51, FC.

7. DM, 3/2/1906; FAF to Robert Amory, 3/8/1938; "Schedule of Wages going into effect March 30, 1908. Drawing-in," PC.
8. L. A. Hackett, "Report on the Boott Mills," 1/9/1911, p. 1, box 45, FC.
9. E. W. Thomas to Dunbar, 4/17/1914, PC.
10. Dr. Ghering to FAF, box 14, FC.
11. FAF Will, box 12, FC.
12. Her sister Mary married Frank Dunbar, and her brother John became Congressman, succeeded by his wife, Edith Nourse; the group of them represented considerable local financial and political power, although not in the same league, economically, as the Ayers.
13. Jacob Rogers to F. F. Ayer, 11/7/1901, PC.
14. Telegrams from F. E. Dunbar to Mrs. Flather, 2/25/1905, 2/18/1905, 2/20/1905; Mary Dunbar to FAF, 2/21/1905, box 14, FC.
15. FAF to Albert E. Lavery, 2/9/20, box 14, FC; the term "employe" in place of the later "employee" remained current in this period.
16. Memorandum, box 1, FC.
17. FAF to Pettee, also legal agreements, letters of negotiations, box 4, FC.
18. Gibb, *Saco-Lowell,* pp. 343, 264–65, 357.
19. FAF to G. F. Steele, 2/9/1895, 3/2/1895, box 4; FAF diary entry, 5/14/1897, box 20, FC.
20. FAF, biographical data for *Lamb's Cotton Industries of the United States,* 1919, box 13, FC.
21. J.R. [Jacob Rogers] to Allie [Alice Rogers Flather], 2/21/1905, box 14, FC; FAF to Mr. McCormick, 12/30/1904, draft, box IH-1, FC.
22. A. C. Thomas to Flather, 6/30/1905, FC.
23. FAF notes, 7/1/92, box 14, FC. While he had invested some $6,000 during the 1890s and seen it grow to around $11,000 by 1904, his substantial salaries had indeed brought him more income. His investments during this period, and thereafter, were conservative, emphasizing utilities, transportation, and major manufacturers, including numerous textile firms. FAF personal account books, box 19, FC.
24. Directors' Reports (hereafter DR) for 8/1929; *Textile World,* 6/21/1924, p. 99, clipping, PC.
25. FAF, Yearbook, 1910, 5/14/1910, box 20, FC.
26. FAF, Yearbook, 1917, 1/29/1917, 2/6/1917, box 21, FC.
27. Ibid., 5/17/1918, box 21, FC.
28. DR, 11/30/1920, FC. The expression meant "idle," as when a belt-driven machine was not running the belt still turned the loose pulley.
29. Letterbook, p. 903, E. W. Thomas to H. Wilson.
30. DR, 11/30/1920, box 32, FC.
31. DR, 8/10/1921, box 32, FC.
32. See also Hall, *Like a Family,* pp. 92–93, 197, 269.
33. Wellington, Sears to FAF, 3/28/1906, PC. His request for advice came as all the more of a surprise since engineering consultants were routinely hired for such assistance when the agent couldn't provide it and, in fact, Lockwood-Greene presented a comparable report less than three months later.
34. FAF, diary, 12/26–7/1907; 1/1/1908, box 20, FC.
35. Ibid, 2/6/1908; *Lowell Courier-Citizen* clipping, 3/13/1908, PC.
36. Wellington, Sears, to Directors, 3/27/1914; "$600,000 Notes to Pay for Mill Improvements," *Lowell Daily Citizen,* 5/16/1914, FC.

37. DM, 1914, p. 11, FC.
38. DM, 3/30/19, PC.
39. DM, 1/29/1914; 3/30/1914; 5/5/1914, PC.
40. "Manufacturing Shares," *American Cotton and Wool Reporter* (n.d.), p. 800, FC.
41. Note, PC.
42. FAF to Albert F. Bemis, 6/22/1914, FC.
43. DM, 7/3/1914, PC.
44. DM, 7/3/1914, p. 19, FC.
45. Hackett, "Report," pp. 14–15, 17, FC.
46. DM, 10/9/1914, PC; FAF, 11/16/1916, diary, box 21, FC.
47. Keyssar, *Out of Work*, p. 265.
48. Valentine, Tead, and Gregg, Industrial Counselors, "The Industrial Audit of the Boott Mills, Lowell, Massachusetts" (1916), box 45, FC, p. 147 (hereafter, the Valentine Audit).
49. Memo, 10/14/1927, PC.
50. Valentine Audit, p. 148.
51. John Rogers Flather (JRF), "Notes," ca. 1929, pp. 13–14, box 59, FC.
52. Scranton, in *Figured Tapestry*, describes the primitive accounting practices in the industry at this time.
53. DR, 7/27/1918, box 32, FC.
54. DR, 12/28/1907, box 40, FC.
55. Valentine Audit, pp. 162, 165–66, FC.
56. Ibid., p. 37.
57. Memo, "Mr. Dillaway, of the Willard Welsh Realty Co., Washington St., Boston, called on Mr. Flather today," 1/31/29, PC.
58. The parallel between this situation and that upriver at the Amoskeag in Manchester, New Hampshire, where the Dumaines, *père* and *fils*, operated for many years is inescapable.
59. JRF, "Notes," 1923–24, box 59, FC.

CHAPTER SEVEN. TECHNOLOGY AND LABOR

1. JRF, "Notes," 9/20/1923, box 59, FC.
2. Copeland, *Cotton Manufacturing*, p. 75.
3. Ibid., p. 82.
4. Ibid., pp. 75, 81–82; DR, 6/30/1906, box 40, FC.
5. See, e.g., Miles, *Lowell*, p. 80.
6. DR, 9/29/1923, FC.
7. DR, 4/26/1913, FC.
8. Treasurer's Report (hereafter TR) for 5/1917, 3/1927, box 33b; DM, 7/27/1906, 7/15/1912, PC.
9. DM, 8/8/1912, PC; TR, 2/1917, box 32, FC.
10. TR, 2/1917, p. 2, box 32, FC; Safford to FAF, p. 5; a substantial portion of the $250,000 Bemis recommended went for turbine replacement in 1919 and 1923.
11. DR, 10/25/1913; Frank E. Rowe, Jr., to FAF, Jr. [sic], 5/26/1931, PC.
12. DR for 11/1927; Navin, *Whitin Machine Works*, p. 513, p. 513n; Gibb, *Saco-Lowell*, p. 558.

13. Francisco Permanyer for Hilaturas Casablancas, S.A., to BM, 3/14/1925, PC.
14. DR for 11/1927 and 2/1928, PC; "Textile Development Corporation Report" (TDCR) 1928, box 45, FC (n.p.).
15. TDCR, pp. 461–71.
16. Ibid., pp. 334–37, 382–88.
17. JRF, 3/21/1924, box 59, FC; Valentine Audit, p. 144.
18. JRF, 3/21/1924; Valentine Audit, p. 144.
19. Valentine Audit, p. 145.
20. F. A. Estes, "Appraisal of Boott Mills, Lowell, Mass.," for Associated Factory Mutual Fire Insurance Companies, Boston, for 1905, 1922, and 1930, PC.
21. Frank E. Rowe, Jr., Sales Engineer, Saco-Lowell Shops, to Frederick A. Flather, Jr. (sic) , 5/26/1931, pp. 3–18, PC.
22. DM, 2/10/1914, PC; TR for 2/1917, box 32, FC; Gross and Wright, HSR, pp. 315, 313; Charles T. Main file, PC.
23. TR for 1/1920, box 32, FC; TR for 9/1930, box 34, FC.
24. Lockwood-Greene to P. D. Saylor, 5/10/1920, PC.
25. See, e.g., Daniel Nelson, *Managers and Workers: Origins of the New Factory System in the United States, 1880–1920* (Madison, 1975), p. 50; Montgomery, *Fall of the House of Labor,* pp. 46ff.
26. DR, 4/26/1913, box 40, FC.
27. Valentine Audit, p. 66.
28. DR, 4/26/1913, FC.
29. Copeland, *Cotton Manufacturing,* p. 33.
30. TDCR, p. 360.
31. Ibid., pp. 358–59.
32. Ibid., pp. 391–94.
33. JRF.
34. TDCR.
35. Valentine Audit, pp. 72–73.
36. TDCR, pp. 537–39; p. 536.
37. Valentine Audit, pp. 92–94.
38. "Less Work for Fixers as well as Weavers," *Cotton Chats,* no. 290 (10/1928), FC.
39. Personal experience has underlined this point.
40. "Payroll Schedule, Sept. 7, 1910," box 79, FC.
41. The seven separate motions were picking, shedding, beating, let-off, take-up, bobbin-changing, and stop motions.
42. TDCR, pp. 516–19.
43. Ibid., pp. 514–15.
44. Hackett, "Report," p. 12.
45. Valentine Audit, p. 33–35.
46. Ibid., pp. 36–38.
47. G. E. Fisher to FF, Jr. (sic), 3/4/1932, PC.
48. Valentine Audit, pp. 40–41.
49. Ibid., pp. 43–44.
50. Ibid., pp. 46–47.
51. TDCR, pp. 134–35.
52. Valentine Audit, p. 40.
53. Ibid., pp. 48–50.

54. Ibid., pp. 50–51.
55. Ibid., pp. 51–52.

CHAPTER EIGHT. PARTICIPANTS AND PROSPECTS

1. Valentine Audit, pp. 99–100.
2. Ibid., p. 105.
3. Ibid.
4. Sheldon to BM, 1/26/1907, p. 2, box 51, FC.
5. Hackett, "Report," pp. 6–8.
6. Ibid., pp., 8–9.
7. Sheldon to BM, 1/26/1907, p. 2.
8. Hackett, "Report," pp. 9–10.
9. Green, *World of the Worker,* p. 5.
10. See, e.g., *American Wool and Cotton Reporter* 17, cited in Chapter 5 above.
11. "Arrangement of Advancement of Wages Week Ending June 8, 1907," box 79, FC.
12. "Payroll Schedule, Sept. 7, 1910," box 79, FC.
13. Montgomery, *House of Labor,* pp. 169–70.
14. DR, 5/10/1916, box 32, FC.
15. Valentine Audit, p. 169.
16. FAF, Notes for Directors Meeting, 5/25/1917, box 32, FC.
17. FAF, Notes for Directors Meeting; DR, 5/25/1917, box 32, FC.
18. DR, 5/25/1917, 11/2/1921, box 32, FC.
19. DR, 5/25/1917, box 32, FC.
20. "Schedule of Numbers of People Employed With Rates Per Hour" and "Schedule of Piece Rate Prices Corrected to Sept. 29, 1919," FC, box 51, pp. 19, 23, 29, 37.
21. DR, 5/24/1920, box 32, FC.
22. Scranton, *Figured Tapestry,* p. 358.
23. TR for 11/1925, box 33b, FC.
24. JRF, box 59, FC.
25. Irving Bernstein, *The Lean Years: A History of the American Worker, 1920–1933* (Boston, 1960), p. 103; Memo, 10/14/1927, PC.
26. Scranton, *Figured Tapestry,* p. 328.
27. DM, 3/2/1906; 2/26/1919, PC.
28. Keyssar, *Out of Work,* pp. 58, 118.
29. Frank W. Hall to FAF, 12/6/07, PC; C. M. Greene, *Holyoke, Mass.,* p. 227.
30. DR, 12/22/1920, 11/24/1920, 4/8/1921, box 33; 5/24/1924, box 59, FC.
31. TR for 6/23, box 33, FC.
32. See, e.g., Green, *World of the Worker,* p. 134.
33. JRF, 4/17/24, FC.
34. Valentine Audit, pp. 170–74.
35. FF to C. E. Holt, 3/6/1926, PC.
36. Valentine Audit, pp. 174, 17.
37. TDCR, p. 98.
38. Valentine Audit, pp. 7–10.
39. Ibid., pp. 10–11.

40. Keyssar, *Out of Work*, p. 288; Parker, *Lowell*, p. 2.
41. Notes, box 15; BM to FF, 3/26/25, box 10, FC.
42. "Investments and Income Record, 1926–1935," box 19; 5/23/1907 document, box 14, FC.
43. Bernstein, *Lean Years*, p. 7; Green, *World of the Worker*, p. 103; Ware, *Early New England Cotton Manufacture*, p. 224 and n.
44. Valentine Audit, pp. 55–56.
45. Ware, *Early New England Cotton Manufacture*, p. 224; and Valentine Audit, pp. 56–59.
46. For a discussion comparing such matters in textile cities, see David J. Goldberg, *A Tale of Three Cities: Labor Organization and Protest in Paterson, Passaic, and Lawrence, 1916–1921* (New Brunswick, 1989), and Gary Gerstle, *Working-Class American: The Politics of Labor in a Textile City, 1914–1960* (Cambridge, England, 1989).
47. E.g., Copeland, *Cotton Manufacturing*, pp. 123, 125; Dunn and Hardy, *Labor and Textiles*, pp. 102, 243 n. 10.
48. Jean-Claude G. Simon, "Textile Workers, Trade Unions, and Politics: Comparative Case Studies, France and the United States, 1885–1914," Ph.D. diss. (Tufts University, 1980), pp. 295–96, 278–79.
49. Mary T. Mulligan, "Epilogue to Lawrence: The 1912 Strike in Lowell, Massachusetts," in Blewett, ed., *Surviving Hard Times*, pp. 82–83; my account relies heavily on this essay.
50. Dexter Phillip Arnold, "'A Row of Bricks': Worker Activism in the Merrimack Valley Textile Industry, 1912–1922," Ph.D. diss. (University of Wisconsin, 1985), p. 391.
51. Valentine Audit, pp. 123–24, 128.
52. Dunn and Hardy, *Labor and Textiles*, pp. 198–99.
53. FAF Notes, 1/20/1917, 3/15/1917, box 54, FC.
54. DR, 4/24/1918, box 32, FC.
55. Edward J. Scollan, "World War I and the 1918 Cotton Textile Strike," in Blewett, ed., *Surviving Hard Times*, pp. 105–14.
56. Leonard E. Tilden, "New England Textile Strike," *Monthly Labor Review* 16 (May 1923): 899–922.
57. DR, 5/24/1920, box 32, FC.
58. Green, *World of the Worker*, p. 120; Dunn and Hardy, *Labor and Textiles*, p. 203; Alice Galenson, *The Migration of the Cotton Textile Industry from New England to the South, 1880–1930* (New York, 1985), p. 115; Bernstein, *Lean Years*, p. 335.
59. TR for 7/1917, 7&8/1919, box 32, FC.
60. Valentine Audit, pp. 89–90.
61. Ibid., pp. 4–6.
62. Brody, *Workers*, pp. 52–53, 56, 57.
63. *Lowell Courier-Citizen*, 8/23/1907, clipping, PC; TR for 2/1917, box 32; TR for 7/30, box 34, FC.
64. [Mrs.] J. M. Gilman to FAF, 10/7/1912, box 76, FC.
65. Working Girls' Club to FAF, 6/7/1911, box 76, FC.
66. Frank P. Bennett, Jr., "The Deflowering of New England," manuscript (ca. 1985), MATH, p. 565.
67. Valentine Audit, pp. 157–58; DR, 10/24/1917, box 32; DR, for 2/1928, box 33, FC.

68. TR for 10/1919, box 32, FC; Bernstein, *Lean Years,* p. 178.
69. Valentine Audit, pp. 160–62; DR, 7/28/1920; BM to Mr. Abbott, 11/28/1923, FC.
70. Bernstein, *Lean Years,* p. 177; Green, *World of the Worker,* pp. 103–4; Daniel Rodgers, *The Work Ethic in Industrial America, 1850–1920* (Chicago, 1978).
71. JRF, 8/2/1923, box 59, FC.
72. JRF, 9/18/1924; TR, 1927, box 33b; JRF to "Mar and Par," 7/11/1928, box 15, FC.
73. Green, *World of the Worker,* p. 102; Bernstein, *Lean Years,* p. 89.
74. Dunn and Hardy, *Labor and Textiles,* p. 75.
75. NECMA became the National Association of Cotton Manufacturers (NACM) in 1906. Flather served as its treasurer in 1930. He and his sons were also members of the American Cotton Manufacturers' Association, the generally southern counterpart of the NACM founded in 1903.
76. TR, 6/1918, box 32, FC; DR, 1/22/1919, box 32, FC.
77. Valentine Audit, pp. 113–14.
78. See, e.g., Lahne, *The Cotton Mill Worker,* p. 258, n. 103.
79. Note, 1/25/1918, PC; TR, 4/1930, box 34, FC.
80. "Mr. Dillaway . . . called," 1/31/1929, PC.
81. Commission for the Protection of Waterways to the Proprietors of the Locks and Canals on Merrimack River, 2/26/1917; Locks and Canals Engineer's Report, 6/26/1922, FC.
82. TR, 3/1920, box 32, FC.
83. Dunn and Hardy, *Labor and Textiles,* pp. 122, 141–42; anon., "The Purpose and Progress of Welfare Work," n.p., PC.
84. TR, 9/1929, box 34, FC.
85. Dunn and Hardy, *Labor and Textiles,* pp. 57, 72.
86. TR, 2/1921, and 12/1922, box 33, FC.
87. Bennett, "Deflowering of New England," p. 565.
88. DR, 1/26/1921, box 32, FC.
89. Arnold, "Row of Bricks," p. 33; TR for 1/1929, p. 34, FC; clipping, PC; TR for 7/1930, box 34, FC.
90. *American Wool and Cotton Reporter,* 19/15/25, clipping, p. 23, PC.
91. Bennett, "Deflowering of New England," p. 566.
92. Dunn and Hardy, *Labor and Textiles,* pp. 16–17.
93. See, e.g., Robert B. Reich, *The Next American Frontier* (New York, 1983), p. 86f, and Scranton, *Figured Tapestry,* p. 449.
94. See Hall, *Like a Family,* for a similar account (e.g., p. 183).
95. Shlakeman, *Chicopee,* p. 228; *American Wool and Cotton Reporter,* 10/18/29, cited in Dunn and Hardy, *Labor and Textiles,* p. 136; see also p. 44.
96. Parker, *Lowell,* p. 152; Dunn and Hardy, *Labor and Textiles,* pp. 182, 186.
97. Copeland, *Cotton Manufacturing,* p. 27; Parker, *Lowell,* p. 104; Charles T. Main, "Report," 1926, PC; JRF noted that Lowell Weaving planned "to move duck looms (famous for fine sails for yachts) to the South" and recorded attending the auction of the Massachusetts Mills; "Notes," 10/9/1923, 4/17/1924, box 59, FC.
98. Lockwood-Greene to P. D. Saylor, 5/10/1920, PC; Main, "Report," pp. 1–5, 10, 19, 28, PC; Parker, *Lowell,* p. 179; Main, "Report," pp. 3–5; A. W. Benoit, Associate, Charles T. Main, Inc. (CTM), "The Growth and Decline of the Number of Cotton Spindles in the United States," Presented at the Regular Meeting of the Textile Division American Society of Mechanical

Engineers, Boston, 10/15/1937, pp. 1–2, 7, PC; Dunn and Hardy, *Labor and Textiles,* pp. 103, 3–5.

99. Benoit, "Growth and Decline," pp. 1–2; *Labor and Textiles,* p. 103; CTM, 1926, p. 10, PC; Main, "Report," p. 19; Benoit, p. 7; Main, "Report," p. 3.
100. Dunn and Hardy, *Labor and Textiles,* p. 47; Main, "Report," 1926, p. 10.
101. Young, *American Cotton Industry,* pp. 23–24.
102. "To the Stockholders of the Tremont and Suffolk Mills," 6/16/1924, box 15, FC; BM "Memo," 2/13/1924, PC; Main, "Report," PC.
103. DM, 12/16/1914, PC; DR, 11/1922, 4/1923, box 32, FC; DM, 1925, PC.
104. TR, 1927, box 33b, FC.
105. TR, 1927; DR, 1929, TR, 1928, box 33b, FC.
106. For discussion of a range of textile mill attitudes, see Hall, *Like a Family.*

CHAPTER NINE. FACTORS AFFECTING SURVIVAL

1. DR, 3/1939. In 1932 Frederick A. Flather stepped down as director, to be replaced by his son, Frederick Flather (FF), who also became assistant manager. John Rogers Flather (JRF) rose to the position of assistant treasurer. The elder Flather remained dominant in the day-to-day operation of the company and a member of the Board of Directors as treasurer; DR, 4/1932.
2. Testimony of FF, PC.
3. "C-283 Complaint," 7/24/1940, PC; TR, 9/32, FC. Customers at times rejected cloth because of production problems: "excess filling bars" in a lot of corduroy, an "off-shade" batch of khaki made for the Marines, the latter a problem caused by the finisher.
4. Stenographic Record, pp. 452, 442.
5. Labor-management file, 1937, p. 2, PC.
6. DR, 8/1939, 9/1939, box 35; 2/1940, 7/1940, box 36; 1/1949, 7/1949, box 37, FC.
7. TR, 1/1940, and 9/1940, box 36, FC.
8. TR, 1/1943, box 36, and 3/1947, box 37, FC.
9. TR, 5/1943, 11/1944, box 36, FC.
10. DR, 1/1942, PC.
11. TR, 5/1943; JRF, "Chronology of Labor Relations," box 78a, p. 50, 4/14/1943; TR, 8/1946, box 37, FC.
12. Insurance Appraisal, 1939; Real Estate file, 1945; Stenographic Record, 1945; DR for 7/1934, PC.
13. Stenographic Record, pp. 711, 26, 73, 132.
14. Stenographic Record, pp. 59, 73, 67, 74, 58, 89, Reinvestment File, 5/17/45, PC.
15. DR, 6/1939, box 35, FC; DM, 8/29/39, *BM v. Tax Board,* pp. 70–71, PC; Memo, 7/19/49; Stenographic Record, pp. 457–58, PC.
16. Stenographic Record, pp. 29, 61, 62, 46, 39, 56, 94, 63, 92, 87–88, 159–60, 145–46, 42, 40, PC.
17. Stenographic Record, pp. 103, 24, 139, 516, 171, PC; see also my "Building on Success," pp. 23–24.
18. Gibb, *Saco-Lowell,* pp. 565, 803 n.
19. DR, 6/1939. box 35, FC; A. T. Norton, *Insurance Appraisal for Arkwright Mutual Insurance Company,* 1939, PC.

20. FAF, Report, 1932, box 26, FC; Robert and Co., Inc., architects and engineers, Atlanta, Georgia, Study of Operations of Merrimack Manufacturing, 6/22/1935, p. 1, PC.
21. TR, 2/1935, box 35, FC; Reinvestment File, 5/17/45; Stenographic Record, p. 45, PC; TR, 5/1947, box 37, FC; DR, 12/1935; Reinvestment File, PC.
22. DR, 3/1939, box 35, FC. Interestingly, the unwillingness to buy new machines contributed to the continuing use of mules. While John Rogers praised their "rounder, evener yarn," most mills had eliminated them by this time.
23. "Machinery and Equipment Purchased Since January 1, 1939," DR, March 1950, box 37, FC.
24. DR, 11/1937.
25. DR, 1940.
26. FF, *DM* for 6/1940, box 36, FC; BM to Kentucky Metal Products Co., 9/3/1954; DR for 11/1949, PC. When, late in 1949, management did purchase a little equipment, it saved "valuable" parts from 50–year-old looms and expected felt pads to have a significant effect on serious vibration problems.
27. Machinery List, 12/1954, box 51, FC.
28. Ralph E. Loper Co., Report to Boott Mills (11/1944), pp. 1–7, box 46, FC.
29. Loper, Report to Boott Mills (11/1944), pp. 1–7, FC.
30. JRF, Record, 8/22/1950, p. 941.
31. Albert Raimond and Associates, Inc., Job Specifications, 4/17/1945, box 47, FC.
32. DR, 12/1944.
33. JRF, Record, 5/18/1943, p. 52.
34. "Survey and Recommendations on Employee Training in the Plant of the BM, Lowell, Mass.," 8/1/1945, pp. 1–3, 8; the following descriptions all come from this source.
35. One of the values of these descriptions is their ability to describe the work involved as a series of steps and make the skills more apparent to those to whom the work is unfamiliar, whether contemporary trainees or modern readers.
36. Copeland, *Cotton Manufacturing*, p. 33.
37. DM, 4/27/1937, PC; TR for 11/1940, 2/1941, 6/1945, 8/1949, box 37, FC; Stenographic Record, pp. 31, 71, 92.
38. Mary H. Blewett, *The Last Generation: Work and Life in the Textile Mills of Lowell, Massachusetts, 1910–1960* (Amherst, 1990), p. 12.
39. Memo, 5/17/45, PC; Eleanor E. Glaessel-Brown, "Mills and Migrants: The Context and Contradictions of Labor Migration," manuscript (Boston, 1985).
40. Dalzell, *Enterprising Elite;* for a similar separation of wages and skill among female workers, see McGaw, *Most Wonderful Machine.*

CHAPTER TEN. OPERATION IN THE DEPRESSION AND WARTIME

1. Bernstein, *Lean Years*, p. 255.
2. TR, 11/1931, box 34, FC; Note, 5/23/1933, PC; financial information for

the period is a compilation of data from company records and the *American Wool and Cotton Reporter;* discussion of general labor trends draws on James Green and others.

3. Green, *World of the Worker,* pp. 140, 150; Irving Bernstein, *The Turbulent Years: A History of the American Worker, 1933–1941* (Boston, 1970), pp. 301–3.
4. James A. Hodges, *New Deal Labor Policy and the Southern Cotton Textile Industry, 1933–1941* (Knoxville, 1986), pp. 87, 99–100, 152.
5. Green, *World of the Worker,* pp. 146–47.
6. TR for 8/1934, DR for 9–11/1934, box 34, FC; DR for 8/1934; 9/5/1934 draft letter to employees, PC; Gary Gerstle, *Working-class Americanism: The Politics of Labor in a Textile City, 1914–1960* (Cambridge, 1989), p. 127; Green, *World of the Worker,* pp. 146–47.
7. DR, 8/1935.
8. DR, 3/1935, box 34, FC.
9. Hodges, *New Deal,* p. 8; Harold H. Young, *Cotton Manufacturing in New England* (Providence, 1928), pp. 29–30; Benoit, "Growth and Decline," p. 4; DR, 6/1938, box 35; TR, 12/1934, box 34, FC; Dunn and Hardy, *Labor and Textiles,* p. 28.
10. Scranton, *Figured Tapestry,* p. 449.
11. Green, *World of the Worker,* p. 167.
12. DR, 5/1940, box 36, FC.
13. Ibid.14. DR, 4/1937, box 34, FC.
15. Paul F. McGouldrick, *New England Textiles in the Nineteenth Century: Profits and Investment* (Cambridge, 1968), p. 210.
16. Labor-Management File, 5/25/37, PC; DR for 5/1937; DR for 6/1937; DR, 11/2/1937, box 34, FC; *Lowell Courier-Citizen* clipping, 6/15/37, PC; FAF to James Rose, 6/21/37, box 15; FW H[all] to FAF, 6/23/37; FW Hall to FAF 6/25/37, box 17, TR, 7/1943, box 36, FC.
17. Bernstein, *Turbulent Years,* pp. 301–2; DR, 11/1937, box 35, FC.
18. DR for 7/1938, 5/1939, box 35, FC; DM, 1/27/37, PC. Newspaper clipping, 1931; 7/14/1933, box 10, FC. Frederick Flathers' financial situation at this time was greatly improved, first by the gift of a house from his father-in-law, and then, through his wife's inheritance at her father's death of the income from $2.5 million (ultimately to be divided among their children).
19. DR, 11/1940, box 36, FC.
20. Green, *World of the Worker,* p. 176.
21. BM to United States Navy, 10/17/1941, JRF Chronology, box 78a, FC.
22. TR, 2/1941, box 36, FC; TR for 9/1942, PC; JRF Chronology, p. 49, box 78a, FC.
23. Green, *World of the Worker,* pp. 166, 176, 181, 183, 194. Dissatisfaction with this state of affairs appeared in the mounting number of work stoppages nationally: 4,000 in 1941, 3,573 in 1943, 4,750 in 1945, 5,000 in 1946, with 116 million person-days lost.
24. DR, 8/1941.
25. DR, 10/1941; spelled variously, "Vaillancourt" appears correct.
26. Ibid.
27. TR for 10/1942, box 36; JRF Chronology, pp. 24, 26, box 78a, FC.
28. JRF Chronology, pp. 14, 18, 21; TR for 10/1942, box 36, FC.
29. JRF Chronology, p. 27, 11/26/1942.
30. TR, 11/1942, box 36; JRF Chronology, pp. 30, 31.

31. JRF Chronology, pp. 34, 45, 49, 50, 51, 56, 59, 62.
32. JRF Chronology, 2/5/1943, p. 38; 10/30/1943, p. 83; 12/29/1943, pp. 91–92; 10/30/1943, p. 83; 7/14/1944, p. 137.
33. Stenographic Record, p. 454; Tax memo, 2/15/1945, PC; JRF Chronology, 2/7/1944, pp. 98–100; 3/9–12/1944, pp. 108–9.
34. JRF Chronology, 3/27/1943.
35. JRF Chronology, 4/22/1942, p. 11; 9/17/1943, pp. 73–76; 7/15/1946, p. 339; 1/31/1946, p. 276; 3/10/1943, p. 43; Memo, W. A. Hawkes and A. W. Gionet to JRF, 2/8/1949, box 47, FC; FAF Memo, 1/10/1944, PC.
36. JRF Chronology, pp. 104, 105, 111, 114, 156–57; Memo, 8/12/1944, PC.
37. Scranton, *Figured Tapestry*, p. 22; JRF, Chronology, pp. 258–61, 175, 167.
38. "Boott Mills' Buildings Are Not Suitable for Post-War Competition in Cotton Manufacturing Nor Present Day Competition Either," 12/15/1944, FC.
39. JRF, Chronology, pp. 111, 119; wherever not otherwise noted, data for this discussion comes from JRF, 12/1945–3/1950; FC.
40. JRF, Chronology, pp. 166, 177, 180, 187, 232, 206–10; TR, 1–2/1945, 5/1945, box 37, FC; JRF cited *Dynamics of Industrial Organization.*
41. JRF, Chronology, pp. 239, 241, 247; TR, 8/1945, box 37, FC.
42. JRF, Chronology, pp. 245, 230–31, 301, 286, 321.
43. JRF, Chronology, pp. 243, 80, 94; Stenographic Record, p. 458; *Boott Mill v. Board of Assessors of Lowell,* "Argument for Appellant," docket nos. 20611, 21748, 22241, Commonwealth of Massachusetts, Appellate Tax Board, p. 53, PC; on p. 58 it is noted that this claim is not supported by the record.
44. Wherever not otherwise noted, data for this discussion comes from JRF, Chronology, 12/1945–3/1950, FC.
45. *BM v. Tax Board,* pp. 18–25, PC; Hall, *Like a Family,* p. 184.
46. Howard Zinn, *A People's History of the United States* (New York, 1980), p. 409; Marc Scott Miller, *The Irony of Victory: World War II and Lowell, Massachusetts* (Urbana, 1988), p. 57; JRF, Chronology, p. 172, 2/20/1945.
47. DR, 11/1945.
48. FAF to Frederick Ayer, 2/17/1937, box 17; TR for 7/1943, box 36; DR, 11/1937, box 35; TR, 3/1947, 12/1946, 10/1944, boxes 36–37, FC; DR for 3/1951, PC; TR, 3/1952, 5/1950, boxes 37–38, FC.
49. Stenographic Record, pp. 68, 136.
50. TR for 5/1947, box 37; DR, 8/1937, box 16, 12/14/1954; box 78a, p. 185, 4/21/1945, FC; interview with Mary Karafelis by Pamela Lehman, 2/26/85; see also Blewett's account in *The Last Generation,* p. 308.
51. DR, 3/1946.
52. TR, 6/1946, 9/1946, 10/1946, box 37, FC; Green, *World of the Worker,* p. 194.
53. DR, 7/1938, 5/1939, box 35; JRF, Chronology, p. 27, 11/26/1942; TR, 3/1945, box 37, FC.
54. *Boott Mills v. City of Lowell,* 3/3/1945, FC, p. 1.

CHAPTER ELEVEN. MANUFACTURING WINDS DOWN

1. TR, 7/1949, 9/1950, 10/1950, box 37; 8/1951, 9/1951, box 38, FC.
2. TR, 10/1950, 6/1949, 12/1950, 11/1950, box 37; 2/1952, box 38; 11/1950, box 37; 7/1953, box 38, FC.

3. TR, 3/1947 and 1/1949, box 37, FC.
4. TR, 3/1949, box 37, FC.
5. TR, 8/1951, box 38; 8/1949, 1/1949, box 37, FC.
6. TR, 1/1947, 7/1948, box 37, FC.
7. Textile Workers Union of America (TWUA) file, 2/26/47, 12/1/47, Wisconsin Historical Society.
8. TR, 8/1947, 11/1947, box 37, FC.
9. FAF to Our Fellow Workers, 12/3/1948, box 74; TR, 1/27/1948, 2/24/1948, box 37, FC; TWUA file, 2/26/1947, 12/1/1947; TR, 11/1947, 1/1948, PC.
10. TR, 2/1952, box 38, 8/1948, 3/1948, 4/1948, box 37, FC.
11. TR, 2/1946, box 37; TR, 4/1946, 5/1947, 4/1948, 7/1949, box 37, FC; "Comparison of Standards and Average Performance 5/1948 and 5/1949, Hourly Paid Jobs," PC; TR, 2/1952, 9/1951, box 38, FC. JRF, Chronology, 10/26/1943, p. 82; 9/2–4/1945, p. 225; 12/3/1945, p. 257; box 38; TR, 2/1948, box 37, FC; TWUA file, 2/26/1947, 12/1/1947; DR for 2/1954, PC.
12. Labor-Management file, 5/1942; JRF to Frank L. Walton, Textile, Clothing and Leather Division, War Production Board, 2/3/1943, PC; TR, 9/1948, box 37; 3/1953, 5/1951, box 38; DR, 9/1939, box 35; TR, 12/1931, box 34, FC; Notes, 9/5/1941, PC; TR, 9/1946, 5/1949, box 37, FC; Boott Mills to Employees, 3/12/1953, PC.
13. TR, 1/1941, 4/1945, 4/1948, "Confidential Supplement" to DR for 5/1948, TR, 5/1948, 5/1949, 6/1949, 9/1949, 2/1950, boxes 36–37, FC.
14. Blewett, *Last Generation,* p. 314.
15. JRF, Chronology, p. 312, 5/1946; TR, 3/1945, box 37; DR, 6/1937, box 35, FC; JRF, "Summary of Union Contract Negotiations," 5/20/1953, Labor Relations file, PC; JRF, Chronology, p. 104; Memo, p. 7, 3/1950, PC; FAF to Cyrus H. McCormick [Sr.], 5/4/36, box 15; TR, 4/1952, box 38; JRF, Chronology, 9/10/1951; JRF and FF to Our Fellow Employees, 4/13/1953, box 74, FC.
16. TR, 10/1949, box 37; TR, 12/1951, box 38, FC; Green, *World of the Worker,* p. 203.
17. DR, Supplement, 12/1949; JRF, Chronology, 12/28/1949.
18. TR, 3/1950, 5/1950, 10/1950, box 37, FC; TR for 2/13/51, PC.
19. TR, 12/1947, 10/1948, 11/1948, 8/1950, 12/1950, box 37, FC.
20. Parker, *Lowell,* p. 138.
21. Memo, 3/1950, PC; TR, 12/1953, box 38, FC; DR for 12/1953, PC; personal communication; JRF Memo, 12/7/1954, PC; "1953 Summary," JRF, 1/8/1954, FC.
22. Memo, 3/1950, pp. 3–5, PC.
23. TR, 9/1950, box 37, FC; "Notes," 10/16/1950, PC.
24. TR, 3/1950, 5/1950, 10/1950, 2/13/1951, box 37, FC.
25. "Straight Time Hourly Earnings for Selected Occupations," 3/1952, FC.
26. TR, 1/1951, 2/1951, 3/1951, 4/1951, 6/1951, 9/1951, 10/1951, 4/1952, 7/1952, 8/1952, 12/1952, box 38, FC.
27. TR, 2/1951, 6/1951, 9/1951, 10/1951, box 38, FC.
28. "Notes" for 12/26/1951, PC; TR, 3/1952, 6/1952, 11/1952, box 38, FC.
29. "Notes" for 12/26/1951, PC; TR, 3/1952, 6/1952, 11/1952, 1 & 2/1953, box 38, FC.
30. TR, 2/1953, box 38, FC.

31. Memo, 3/20/1935, PC; TR, 8/1947, box 37; TR, 12/1953, box 38; Note, 11/30/1945, PC; TWUA manuscripts 129A, 2A, 11/10/1949, 10/22/1952, 1/14/1949, 11/15/1951.
32. TR, 1/1953, 4/1953, 5/1953, 6/1953, 7/1953, 9/1953, 10/1953, box 38, FC.
33. TR, 2/1951, 5/1952, 2/1953, 1/1954, box 38, FC.
34. JRF, Chronology, 10/26/1943, p. 82; 9/1922–4/1945, p. 225; 12/3/1945, p. 257; TR, 2/48, box 37, FC; TWUA file, 2/26/1947, 12/1/1947; *DR* for 2/1954, PC.
35. Such a career path existed in Columbian textiles; see Glaesser-Brown, *Mills and Migrants.*
36. TR, 4/1953, box 38, FC.
37. TR, 3/1953, 4/1953, 8/1953, 2/1954, 5/1954, box 38, FC.
38. TR, 7/1953, 11/1953, 12/1953, PC.
39. BM to John Smith, 4/16/1954.
40. TR, 1/1954, 2/1954, 3/1954, 5/1954, 6/1954, 10/1954, box 38, FC.

CHAPTER TWELVE. TWO POINTS OF VIEW

1. TR, 12/1940, box 36; FAF to John S. Lawrence, 10/15/1935, box 17; DR, 11/1937, box 35, FC.
2. Sketch letters, FAF, 6/4/1936, box 17; DR, 1/1939, box 35; TR, 12/1948, box 37, FC.
3. DR, 11/2/1937, PC; DR, 9/1938, box 35; TR, 9/1950, box 37; DR, 6/1939, box 35, FC.
4. DR, 1951, PC; TR, 12/1934, box 34, FC; "Increasing Income and Decreasing Expenditures for the Purpose of Reducing Losses Can Hardly Be a Quick Job," PC; DR, 12/1938, box 35; TR, 8/1953, box 38, FC.
5. Sanderson and Porter, *Report on Merrimack Manufacturing,* p. 40; Stenographic Record, p. 16, p. 427; various letters, PC.
6. Management "Notes," 5/26/1943; FF to Edmund W. Sanderson, Great Atlantic and Pacific Tea Co., 7/29/1931, PC; TR, 9/1952, box 38, FC.
7. FF letter, Labor-Management 1930s file, 2/25/1935, PC; JRF, Chronology, p. 57, 6/7/1943; TR, 3/1943, 11/1942, box 36, FC; Tax Memo, 2/15/1945; "The Boott Case Against a Wage Advance," PC.
8. DR, 7/1942, PC; JRF, Chronology, p. 139, 7/21/1944, FC; Stenographic Record, p. 464; Flathers to Employees, 5/7/1953, PC; TR, 2/1953, box 38, FC.
9. JRF, Chronology, 4/1/1947, box 78, FC.
10. DR, 4/1939, box 35, FC.
11. JRF Memo, 12/7/54.
12. Bernstein, *Lean Years,* pp. 299–300.
13. TR, 3/1952, 11/1953, box 38.
14. Brody, *Workers,* p. 127; Louie Vergados interview, 10/13/1982, University of Lowell Special Collections, p. 66.
15. Harry Dickenson, interviewed by Judith K. Dunning, 9/17/1980, p. 8, Lowell National Historic Park.
16. Dickenson, interview, pp. 12–15, 22–23.

17. JRF, Chronology, p. 65, 7/3/1943; TR, 2/1948, box 37, FC; *BM v. Tax Board*, pp. 71–72, PC.
18. Joe Higgenbottom to FAF, 2/18/1955; FAF to Joseph J. Higgenbottom, 2/21/1955, PC.
19. Bennett, "Deflowering of New England," pp. 566, 610.
20. See, e.g., *American Wool and Cotton Reporter*, 2/18/1932; Miller, *Irony of Victory*, p. 144, and the Merchants National Bank of Boston: Diversification of Investments, 11/19/1958; Tax Appeal, p. 31.

POSTSCRIPT

1. Mark Reutter, *Sparrows Point: Making Steel—The Rise and Ruin of American Industrial Might* (New York, 1988), p. 265. Dalzell, *Enterprising Elite*; David Bensman and Roberta Lynch, *Rusted Dreams: Hard Times in a Steel Community* (New York, 1987), p. 88.
2. Scranton, *Figured Tapestry*, pp. 12, 160, 388.
3. *NECMA* 13 (1872): 8; I am indebted to Patrick Malone for this citation.
4. Brody, *Workers in Industrial America*.
5. Lockwood-Greene to P. D. Saylor, 5/10/1920, PC.
6. Reich, *Next American Frontier*, p. 9.
7. Webber, *Manual of Power*, p. 97; various records, FC.
8. Reich, *Next American Frontier*, p. 82; Reutter, *Sparrows Point*, p. 236.
9. Michael L. Dertovyos et al., *Made in America: Regaining the Productive Edge* (Cambridge, 1989), pp. 41, 82–83, 295; Stephen Cohen and John Zysman, *Manufacturing Matters: The Myth of the Post-Industrial Economy* (New York, 1987), p. 199; works cited represent only a sampling of the literature on deindustrialization, its consequences, and its alternatives.
10. William J. Abernathy, Kim B. Clark, and Alan M. Kantrow, *Industrial Renaissance: Producing a Competitive Future for America* (New York, 1983), p. 11.
11. Robert B. Reich, *The Work of Nations: Preparing Ourselves for Twenty-first - Century Capitalism* (New York, 1990); Robert Kuttner, *The End of Laissez-Faire: National Purpose and the Global Economy after the Cold War* (New York, 1991); Shoshana Zuboff, *In the Age of the Smart Machine: The Future of Work and Power* (New York, 1988); John Case, *From the Ground Up: The Resurgence of American Entrepreneurship* (New York, 1992); Richard Rosecrance, *America's Economic Resurgence: A Bold New Strategy* (New York, 1990); David Halberstam, *The Next Century* (New York, 1991); Michael J. Piore and Charles F. Sabel, *The Second Industrial Divide: Possibilities for Prosperity* (New York, 1984), in addition to those cited here, are a few of the ever-growing list of players. The sources cited represent simply a sampling of this ever-changing field.
12. Doug Bailey, "State Needs Lead Industry, Says Syron," *Boston Globe*, 4/3/1990, pp. 51–56.
13. Kenngott, *Record of a City*, p. 163.
14. Box 14, FC.
15. Thomas A. Kochan, Harry C. Katz, and Robert B. McKersie, *The Transformation of American Industrial Relations* (New York, 1986), p. 227.
16. Paul C. Weiler, *Governing the Workplace: The Future of Labor and Employment Law* (Cambridge, 1990), p. 307.

17. Barbara Garson, *The Electronic Sweatshop: How Computers Are Transforming the Office of the Future into the Factory of the Past* (New York, 1988), p. 70.
18. Kuttner, *Laissez-Faire*, p. 263.
19. Ibid., pp. 273–74.
20. Kochan, Katz, and McKersie, *Industrial Relations*, p. 231.
21. See, e.g., Charles Stein, "Fraying of the Blue-Collar Worker," *Boston Globe*, 4/28/1992, pp. 1, 20; Louis Uchitelle, "Blue-Collar Compromises in Pursuit of Job Security," *New York Times*, 4/19/1992, pp. 1, 18.

INDEX

Books in the New Series

The American Railroad Passenger Car, New Series, no. 1, by John H. White, Jr.

Neptune's Gift: A History of Common Salt, New Series, no. 2, by Robert P. Multhauf

Electricity before Nationalisation: A Study of the Development of the Electricity Supply Industry in Britain to 1948, New Series no. 3, by Leslie Hannah

Alexander Holley and the Makers of Steel, New Series no. 4, by Jeanne McHugh

The Origins of the Turbojet Revolution, New Series no. 5, by Edward W. Constant II (Dexter Prize, 1982)

Engineers, Managers, and Politicians: The First Fifteen Years of Nationalised Electricity Supply in Britain, New Series no. 6, by Leslie Hannah

Stronger Than a Hundred Men: A History of the Vertical Water Wheel, New Series, no. 7, by Terry S. Reynolds

Authority, Liberty, and Automatic Machinery in Early Modern Europe, New Series, no. 8, by Otto Mayr

Inventing American Broadcasting, 1899–1922, New Series, no. 9, by Susan J. Douglas

Edison and the Business of Innovation, New Series, no. 10, by Andre Millard

What Engineers Know and How They Know It: Analytical Studies from Aeronautical History, New Series, no. 11, by Walter G. Vincenti

Alexanderson: Pioneer in American Electrical Engineering, New Series, no. 12, by James E. Brittain

Steinmetz: Engineer and Socialist, New Series, no. 13, by Ronald R. Kline

From Machine Shop to Industrial Laboratory: Telegraphy and the Changing Context of American Invention, 1830–1920, New Series, no. 14, by Paul Israel

The Course of Industrial Decline: The Boott Cotton Mills of Lowell, Massachusetts, 1835–1955, New Series, no. 15, by Laurence F. Gross